CONFRONTING CREATIONISM: DEFENDING DARWIN

Australian Studies in Biological Sciences Series

In preparation

Australian Institute of Biology

The Australian Institute of Biology (AIB) was formed as an independent body in 1986 after a number of years as a branch of the Institute of Biology based in the United Kingdom. The AIB aims to represent the biology profession in Australia to all levels of government and the media for the purpose of promoting education and research in biology. It also arranges national and local meetings, often in association with scientific societies. Besides providing the general editors for Australian Studies in Biology series, the Institute publishes the *AIB Newsletter*.

CONFRONTING CREATIONISM: DEFENDING DARWIN

edited by

D.R. Selkirk

and

F.J. Burrows

THE NEW SOUTH WALES UNIVERSITY PRESS

in association with the
AUSTRALIAN INSTITUTE OF BIOLOGY

Published by
NEW SOUTH WALES UNIVERSITY PRESS
PO Box 1 Kensington NSW Australia 2033
Telephone (02) 697 3403

First published in 1987
Reprinted with minor corrections 1988

National Library of Australia
Cataloguing-in-Publication entry:

Confronting creationism: Defending Darwin

   Bibliography.
   Includes index.
   ISBN 0 86840 178 1.

   1. Evolution. 2. Evolution — Religious aspects.
   3. Creationism. I. Selkirk, D.R. II. Burrows, F. J.
   (Frank J.).

575

Available in North America through:
International Specialized Book Services
5602 N.E. Hassalo Street
Portland Oregon 97213-3640
United States of America

Printed and bound in Australia by
Southwood Press Pty Limited, Marrickville, 2204

CONTENTS

THE GEOLOGICAL TIME SCALE

	Era	Period	Epoch	Age in years
PHANEROZOIC	CENOZOIC	Quaternary	Recent	0–10 000
			Pleistocene	10 000–3 000 000
		Tertiary	Pliocene	3–5 000 000
			Miocene	5–22 500 000
			Oligocene	22.5–37 500 000
			Eocene & Paleocene	37.5–65 000 000
	MESOZOIC	Cretaceous		65–135 000 000
		Jurassic		135–192 000 000
		Triassic		192–225 000 000
	PALEOZOIC	Permian		225–280 000 000
		Carboniferous		280–345 000 000
		Devonian		345–395 000 000
		Silurian		395–435 000 000
		Ordovician		435–500 000 000
		Cambrian		500–570 000 000
PRECAMBRIAN	PROTEROZOIC	Adelaidean		570–1 400 000 000
		Carpentarian		1400–1 800 000 000
		Lower Proterozoic		1800–2 300 000 000
	ARCHAEAN			2 300 000 000 – 4 500 000 000

after Harland et al 1982

GENERAL PREFACE TO THE SERIES

In 1966, as a response to the rapid and important progress being made in many areas of the biological sciences, the Institute of Biology began publication of a series of booklets, Studies in Biology, dealing with specific biological topics. The series proved tremendously successful, and of great value to students for whom textbooks have tended to become decreasingly able to span the growing breadth of their subject.

Twenty years later, in its first year of independent existence, the Australian Institute of Biology (with the New South Wales University Press) has begun publication of an Australian Studies in Biology. Books in this series will cover topics from a wide range of the biological sciences and their applications. Each will address a topic of particular concern to Australian biologists, and they will be authored by specialists with extensive Australian experience. They will therefore prove especially useful in providing an Australian perspective, supplementing material in textbooks which have often been prepared primarily for students in other countries.

Australian Studies in Biology are written with undergraduate students particularly in mind. However they should also be of great value to biology teachers at both secondary and tertiary level, and to all people having an interest in aspects of the biological sciences in this country.

J.M.B. Smith
Chairman, Publications Committee, Australian Institute of Biology;
Department of Geography and Planning, University of New England

FOREWORD

Religious faith and science have no quarrel. Both are ways of trying to understand the universe and our place in it. Should either faith or science masquerade as the other the result is confusion.

To all concerned scientists a worrying source of confusion is that created in the minds of many, particularly schoolchildren making their first tentative steps into the world of science, by the so-called 'scientific' teachings of a fundamentalist Christian faith known as Creation 'Science'. These teachings use the word of the Bible to challenge the conclusions of scientists based on observation and experimentation. They have been introduced as a legitimate part of the secondary school science syllabus in Queensland, are constantly expounded in the media, and pressure to include them in other science syllabuses is being brought to bear on public and politicians alike.

This book is an expression of concern at these manoeuvrings. It is an expansion of the proceedings of a symposium convened to address specific issues raised by Creation 'Science' and to state firmly and clearly the scientific evidence for the origin of the universe and the evolution of man. The symposium — 'In Defence of Science' — was convened jointly by the Institute of Biology in Australia, the Linnean Society of New South Wales, the Royal Zoological Society of Australia and the Australian Museum Society. The speakers were Mr Ronald Strahan, Executive Officer, National Photographic Index of Australian Wildlife; Professor Ronald Brown, Department of Chemistry, Monash University; Dr Alex Ritchie, Head of the Department of Palaeontology, the Australian Museum; Associate Professor Michael Archer, School of Zoology, University of New South Wales; and Dr David Briscoe, School of Biological Sciences, Macquarie University. Several of these speakers have strongly restated the evidence for evolution in other arenas.

The book is aimed primarily at high school teachers but written so that it will be accessible to senior science students. It provides not only a readable account of the scientific evidence for evolution but also answers some of the more frequently stated claims of Creation 'Science' which contradict the sciences of astronomy, geology, palaeontology and biology and also contradict religion. Aspects of all of these are addressed in this book.

We are indebted to all contributors, particularly Associate Professor

Michael Archer who greatly expanded the substance of his talk to clarify the major evidence on which the evolutionary model is based and to expose the absence of science in the claims of creation 'Science'. Professor Archer also provided the partially annotated bibliography at the end of Chapter 7 and much of the very comprehensive list of references. Our thanks to Professor Ian Falconer, Department of Biochemistry, Microbiology and Nutrition, University of New England, who contributed the final chapter in which he clearly demonstrates the lack of conflict between true Christianity and science.

D.R. Selkirk and F.J. Burrows (Editors)
School of Biological Sciences, Macquarie University

CHAPTER ONE
THE CREATIONISM CRUSADE

Ronald Strahan

WHY THIS SYMPOSIUM WAS CONVENED

Biology is a well-established and largely self-consistent science and the concept of organic evolution is one of its basic unifying principles applicable from the level of ecology to that of molecular biology. No competent biologist would suggest that our understanding of the processes of evolution is complete or correct in every detail but neither would he or she abandon the concept unless a better, more comprehensive theory were offered in its place: rejection of evolution would fracture biological science into a jumble of unrelated data. It therefore seems ridiculous as we approach the close of the twentieth century to convene a public symposium to *defend* the theory of evolution — as pointlessly redundant as defending the spheroid shape of the Earth or the electronic theory of valency. Yet biologists have been forced into this defensive, or counter-offensive, position in response to an evangelical religious movement which claims that the theory of evolution is not only false but diabolically inspired. In its place we are offered the account of Divine Creation given in the opening chapters of the Book of Genesis.

Few scientists are prepared to devote much time or effort to combating anti-science. We look with sadness or scorn upon the proponents of astrology, palmistry, iridology, water-divining, pyramidology, scientology, naturopathy, a flat Earth, flying saucer visitations, and comparable fads, fallacies or fantasies. Why then are we upset by Creationist cranks? In a democratic pluralist society nobody can object to the right of individuals or groups to adhere to any set of beliefs, particularly when these are of a religious nature and, in the name of religious and educational freedom, we permit children to be indoctrinated with the extreme views of their parents in special schools. This symposium has not been convened to object to the rights of people to be silly but to oppose their claim to inflict their silliness on everybody else through government schools. The Australian Institute of Biology, the Linnean Society of New South Wales, the Royal Zoological Society of New South Wales and the Australian Museum Society are concerned to oppose attempts to have Creationism included in the science syllabuses of secular government schools because this would destroy science as we know it. As admitted by one Australian creationist: 'If we reject the evolutionary framework then we must also be prepared to have a revolutionary appraisal of the entire scientific tradition.' (Scott, 1980).

THE CREATIONIST DOGMA

Creationists rest their entire case on a belief in the inerrancy of the Bible, particularly the accounts of Creation and a universal Flood given in the Book of Genesis. The salient points may be summarised as in Table 1.1.

Despite some internal contradictions, this account is taken to be the literal truth. The days referred to are of twenty-four hours, not metaphors for longer periods. The age of the Earth and the universe, calculated from the (incomplete) chronology of the Bible, is said to be between 6000 and 20 000 years and, whatever the age of mankind, *this is only four days less than that of the universe itself.* The Earth was formed before the sun, moon and the stars. Land plants came into existence prior to any aquatic organisms. Birds preceded reptiles. The first human female was cloned from the body of the first human male. All animals were herbivorous until Adam and Eve fell into sin. Noah's Flood (around 2200 BC) is responsible for almost all the sedimentary rocks and all fossils are therefore of virtually the same age. The intellectual contortions required to defend these views are ludicrous but nevertheless comprise a· large body of pseudoscientific literature.

Yet another tenet of the Creationist credo is that all the different 'kinds' of animals came into existence in the fifth and sixth days of Creation and that none has appeared since. The definition of 'kind' has a comfortable elasticity, first being interpreted to mean species but more recently expanded by some Creationists to include higher taxonomic levels. In the 1981 court case in Arkansas (referred to later in this chapter) the following passage occurred in the examination of a Creationist witness (see Gallant, 1984).

Q. How many originally created kinds were there?
A. Let's say 10 000 plus or minus a thousand.
Q. Some creationists believe kinds to be synonymous with species, some with genera, some with families, some with orders, don't they?
A. The scientists with whom I am working — well — it tends more towards the family. But it may go to order in some cases.
Q. You have been studying turtles for many years, haven't you?
A. Yes.
Q. Is the turtle an originally created kind?
A. I'm working on that.
Q. Are all turtles within the same created kind?
A. That's what I'm working on.

Just how the witness could hope to decide, on the basis of research, what constitutes a Biblical 'kind' is puzzling particularly so because many of his fellow believers maintain that Creation must forever remain mysterious. 'We do not know how the Creator created, what processes He used, *for He used processes which are not now operating anywhere in the natural universe.*' (Gish, 1974a).

Putting scientific investigators of origins even more firmly in their place, four leading Australian proponents of creationism state: 'What we believe about where we came from is a serious matter indeed...In the end it becomes a choice between faith in the words of a God who was there or a belief in the words of men who were not.' (Snelling *et al.*, 1983).

Elsewhere, one of these authors makes this point even more precisely: '...we base our understanding upon a book which claims to be the Word of One who knows everything there is to know about what happened.' (Ham, 1983).

A little further down the same page he makes a very strange generalisation: 'When we discuss creation/evolution, we are talking about beliefs: i.e. religion. The controversy is not religion versus science, it is religion versus religion, and the science of one religion versus the science of another.' (Ham, 1983).

This is nonsense but *intentional* nonsense — wilful obfuscation that exemplifies the current strategy of the creationist movement. To understand its significance we must review the history of creationism.

THE OLD TESTAMENT, BIBLICAL CRITICISM AND DARWINISM

Christianity began as a reformist movement within the Jewish religion and, although it soon became separate, many Jewish writings were retained as part of its holy scriptures. Among these is the Book of Genesis which almost certainly represents ancient orally transmitted myths and legends, some of which are derived from pre-Jewish Babylonian and Sumerian sources (see Chapter 8 of this volume).

The version of Genesis accepted by Christians is, like the rest of the Old Testament, translated from the oldest surviving Hebrew version, known as the Massoretic text. This was written down in the ninth and tenth centuries AD but is certainly derived from earlier versions. The partial Greek and Samaritan translations of the Old Testament dating back to the second century AD show that several Hebrew versions were in existence at that time *and that these versions differed from each other and from the Massoretic text*. It requires a considerable act of faith to regard a seventeenth-century English translation of a ninth-to-tenth-century Hebrew document as an authoritative account of the origins of the universe.

Yet this was the case over a very long period. When Christian bishops eventually came to an agreement on what documents should be included in the Bible they effectively authorised this collection as 'the word of God' — written by humans but divinely inspired and therefore true. However, as

early as the third century AD the Christian philosopher Origen recognised
that there were passages in the Old Testament which could not be literally
true and others that were in conflict with Christian doctrine. His solution
was to say that where the Bible is apparently untrue or immoral it hides a
deeper allegorical meaning — an exegesis that has proved very convenient
to liberal Christians ever since.

In fact Christians specifically rejected much of the teaching of the Old
Testament and it is largely because of this rejection that Christianity differs
from the Jewish religion.

Nevertheless, the Book of Genesis remains an integral part of the
Christian scriptures because the story of Adam and Eve explains the origin
of the sins for which Jesus is said to have atoned with his death. The fact
that it is based on the concept of a flat Earth covered by a hemispherical
heaven seems not to have concerned mediaeval scholars and, by the twelfth
and thirteenth centuries, educated Christians were able to regard this as
allegorical. Genesis certainly implies that the Earth is the centre of the
universe but by the end of the seventeenth century the Church was able to
accept a heliocentric system of planets including the Earth. This effectively
resolved conflict between science and the Bible for several centuries. The
comforting view that the universe, the Earth and all the organisms on it had
been created to serve humans was generally held to by theologians and
scientists alike.

Science is blamed for many things but it must be recognised that it was
within the Church that criticism was nibbling away at the authority of the
Bible. Dedicated Christian scholars found that the scriptures were not what
they claimed to be — that they were full of human error and interpolations.
By the nineteenth century, Biblical criticism was a respectable scholarly
pursuit and, as the documents were demonstrated more and more to be the
work of biased and fallible authors, the view began to arise amongst many
committed Christians that the Bible had to be regarded more as inspir-
ational than inspired. Referred to as 'modernism', this approach to the
scriptures was initially restricted to scholars within the orthodox churches
but its influence gradually began to be expressed from pulpits — to the
dismay of the more evangelical and nonconformist Christians. In general it
can be said that nonconformists, despite their concern for the individual
conscience, have not taken a questioning intellectual approach to the
scriptures and, to make a further generalisation, that the greater their
nonconformity the greater their belief in the literal truth of the Bible. This
is a highly simplified description of the situation but the point I wish to
make is that the intense disagreement over interpretation of the Bible which
reached a crisis in the mid-nineteenth century was not the outcome of a
clash between science and religion but between theologians themselves.

At this time, however, scientific ideas were again coming into conflict
with a literal interpretation of Genesis. The emerging discipline of geology
could not be accommodated within a world that was created in 4004 BC.

Nor could the record of the rocks be reconciled with Noah's Flood. Nor was the fossil record consonant with the simultaneous Creation of all species of living things. However, an accommodation seemed possible if the Genesis account of the 'days' of Creation were regarded as allegorical and the Flood as an exaggerated account of a local occurrence.

It was into this situation that Darwin's bombshell was thrown. His was not the first scientific theory of evolution but it was the most comprehensive one ever put forward, linking a large number of observations and postulating a mechanism by which evolution could proceed.

In a matter of years it was accepted by most biologists as a unifying hypothesis even though, as a few perceptive critics pointed out, its genetical basis was flawed. Most churchmen reacted with kneejerk opposition but it is interesting to note that, within a few decades, the orthodox churches reached an accommodation with the *fact* of evolution if not with Darwin's suggested mechanism. The conflict appeared to have been largely resolved by the end of the nineteenth century.

THE FUNDAMENTALIST BACKLASH

I have said that the conflict *appeared* to have been resolved but the nonconformist evangelical right wing of Christianity regarded the rapprochement as betrayal — an unholy alliance between modernist critics within the Church and atheistic mechanists outside. Strangely, they made little organised public response until the second decade of the twentieth century.

In 1910 the General Assembly of the Presbyterian Church replied to the modernists by enunciating five fundamental doctrines of Christianity, including the literal truth of the Bible. In the USA between 1910 and 1915, twelve significant booklets under the general title *The Fundamentals* were published in opposition to modernism and these served as the basis of a body of belief known as 'fundamentalism', claiming to encapsulate the views of traditional protestant Christianity. Again it should be noted that fundamentalists were initially concerned to oppose what they saw as backsliding within the orthodox churches and that the crusade against evolution did not begin until after World War I. My concern, therefore, is with the anti-evolution activities of fundamentalists since that time.

Organised fundamentalism began as and remains a loose linkage between churches and associations, often with shared membership. Among the founding bodies were the Moody Bible Institute, the World's Christian Fundamentals Association, the Anti-Evolution League of America, the Bible Crusaders of America and the Flying Fundamentalists. The strength of these bodies came largely from the southern and middle-western regions of the USA — the Bible Belt — where, between 1921 and 1929, fundamentalists exerted such political pressure that the teaching of evolution was banned by law in seven states.

Under one of these laws a charge was brought in 1925 against a high school teacher, John Scopes, accused of teaching evolution in the town of Dayton, Ohio. The trial was a test case prosecuted by the famous William Jennings Bryan and defended by the equally famous Clarence Darrow. In a courtroom drama that was reported around the world, Scopes was convicted of teaching 'the descent of mankind from lower animals' and fined $100. As a 'test' the trial was a success but the ridicule to which the prosecution was exposed was so great that no further cases were ever brought in Ohio or in any other state.

Despite this reverse, the fundamentalists had not been put out of action. Working on and in state departments of education, local school boards and parent organisations they were effective in banning the use of high school biology textbooks that included serious treatment of evolution. In order to maintain their sales, publishers began to pare down or delete references to evolution in their high school textbooks and so, irrespective of whether a state or school *wished* to teach evolution, the requisite books were hard to find. From the 1930s to the 1950s, over the greater part of the USA most high school pupils learnt little more about evolution than that it was a 'theory' which some people believed in. Given this gap in their scientific education, it is not surprising that the US population today is remarkably receptive to anti-evolution propaganda.

The situation began to change after the first Sputnik circled the Earth in 1957. Regarding this as evidence that the USSR was ahead of the USA in science, the Federal government initiated a number of urgent curriculum development programs in science and mathematics, aimed at improving the teaching of these subjects in the state public schools. One of these was the Biological Sciences Curriculum Study (BSCS) which was designed to introduce students to modern biology and, of necessity, the three BSCS textbooks produced in 1963 and 1964 had a strong evolutionary component. A parallel program, 'Man, a Course of Study' (MACOS), linked biology and the social sciences in a course for upper primary pupils. First published in 1970 this also had an integral component of evolution.

Both BSCS and MACOS, but more particularly the latter, aroused strong opposition from fundamentalists who were able to ban the use of these texts in many schools. Nevertheless, the high quality of the texts and their federal imprimatur led to their wide acceptance: the Federal government had not set out to attack the fundamentalists but had outflanked them at the State level. Most educationists agreed that science teaching had to be improved and that an upgrading of biology teaching required that the unifying concept of evolution be properly presented.

CREATION 'SCIENCE'

The fundamentalist response was utterly dishonest but brilliant: if *science* was to be the issue they would provide their own version of it. In 1963 a group of ten fundamentalists founded the Creation Research Society, dedicated to proving that Genesis is not only compatible with science but that it provides *a more scientific account* of origins than present-day cosmology, geology and biology. The Creation Research Society has a sister institution, the Institute for Creation Research, and to be a full member of either body requires the possession of a post-graduate degree. It is not surprising that the membership includes very few biologists or that none of these has any professional standing since, before being accepted, an applicant must sign a form affirming belief that:

1. The Bible is the written word of God...this means that the account of origins in Genesis is a factual presentation of simple historical truths.

2. All basic types of living things, including man, were made by direct creative acts of God during Creation Week as described in Genesis. Whatever biological changes have occurred since Creation have accomplished only changes within the original created kinds.

3. The great Flood...was an historical event, worldwide in its extent and effect.

4. ...The account of the special creation of Adam and Eve as one man and one woman and their subsequent fall into sin is the basis for our belief in the necessity for a Savior for all mankind... (Gallant, 1984b).

It is self-evident that nobody who makes such a commitment of faith can claim to take a scientific view of questions relating to origins. In fact, the affirmation denies the possibility of questions and a Creation 'Scientist' can do no more than seek evidence in favour of his absolutely settled beliefs and reject all evidence that conflicts with it. *This is the absolute opposite of a scientific approach.*

THE NATURE OF SCIENCE

What we call science comprises a wide range of methodologies that is difficult to summarise. Here, however, it is sufficient to indicate some of the differences between scientific explanation and the sort of religious explanation that is exemplified by the Book of Genesis.

Truth and hypothesis

Genesis offers an explanation which is claimed to be absolutely true. A scientific explanation, on the other hand, is always tentative, for which reason we refer to it as an hypothesis or theory. (A scientific 'law' is not an explanation, merely a description of an apparently constant relationship between certain variables.)

Falsifiability

A scientific hypothesis must be so constructed that in principle it could be shown to be partially or completely false. Most scientists would claim that the account of origins given in Genesis has already been shown to be false but, as presented by Creation 'Scientists', the hypothesis is not open to question. It therefore lies outside the realm of science.

When pushed, Creation 'Scientists' admit that they are dogmatic and that their faith is immovable but they then claim that scientists are equally dogmatic (or religious) in their support of the hypothesis of evolution. This is the significance of an earlier quotation: 'The controversy is not religion versus science, it is religion versus religion . . .' (Ham, 1983).

This is wilful distortion. The theory of evolution is accepted by scientists because it provides an explanation for a wide range of observations (or 'facts') in the geological and biological sciences but it *could* be shown to be invalid, or in need of radical revision, by many conceivable observations. The demonstration of human fossils in Carboniferous rocks, of vertebrates with three pairs of limbs (as in angels), or of a group of frogs with three ossicles in the inner ear would certainly lead biologists to an agonising reappraisal. Since it would be easy to extend the list of conceivable falsifications to thousands of items, the fundamentalist claim that the evolutionary hypothesis is religious is mischievous.

Testability

In addition to passing the test of falsifiability a scientific hypothesis must be testable in the positive sense. It must be shown to be consistent with all phenomena in the area that it claims to embrace and also to lead to testable predictions. In the physical sciences tests are usually applied to predictions within a very short time frame. Some fields of biological science (e.g. physiology, embryology, genetics) fall into a similar category but most questions of organic and inorganic evolution postulate such immense periods of time that they are usually tested in retrospect: i.e. if the theory is valid a particular state of things should have existed in a particular place at a particular time in the past. In this sense the theory of evolution has been well tested.

It is pertinent here to explain a point which is misunderstood by many people (including scientists). When an hypothesis has passed what we regard as a sufficient variety of tests we say that it has been *proved*. But 'proof' is merely another word for 'test' (as in 'the proof (test) of a pudding is in the eating' or 'the exception proves (tests) the rule'). To say that a theory has been 'proved' does not mean that it is true in an absolute sense — only that it has not been falsified by any of the tests to which it has been subjected. It requires some humility to recognise that however inclusive and powerful a scientific hypothesis may be, it will cease to be a *scientific* explanation as soon as we regard it as absolutely true.

Here again we see the gulf between Creation 'Science' and actual science. Creation 'Scientists' do not say, 'Our creation hypothesis is open to testing in respect of all its implications'. They simply say that their postulates are true. For this reason their beliefs lie outside the framework of science.

Corrigibility

The history of science demonstrates that some scientific theories persist for only a short period before being falsified. Others are demonstrated to refer only to a limited range of phenomena and become subsumed by wider, more unifying explanations. Science is a communal activity which proceeds by severe internal criticism, rigorous investigation and occasional flashes of insight. By training ourselves to limit our respect for authority and to use doubt constructively, we continually put our theories to the test and subject them to rejection or refinement. Science proceeds by self-correction but Creation 'Science' is, by its very nature, incorrigible. Let me return once more to an earlier quotation and examine its significance: 'The controversy is not religion versus science, it is religion versus religion, and the science of one religion versus the science of another.' (Ham, 1983).

This is simultaneously a puerile debating point (the falsity of which has already been exposed) and a fall-back position for a constitutional legal challenge. If the teaching of Creation in government schools is judged to be contrary to prohibitions on the teaching of religion therein, Creationists argue that evolution is itself a 'religion' and therefore should also be banned. One's mind boggles at such casuistry.

SUCCESSES OF CREATION 'SCIENCE'

The strategy of Creation 'Scientists' is to claim that they have an explanation of origins which is just as scientific as that offered by biologists and that, as a matter of fair play and educational honesty, their 'scientific' views should be presented with equal emphasis wherever evolution is taught. In this way they avoid the US constitutional prohibition on teaching *religion* in public schools. However ridiculous this may seem, it works, particularly in alliance with the resurgent right-wing political movement in the USA and not least with the current President. In the course of his second election campaign he remarked of evolution: 'Well, it is a theory, it is a scientific theory only and it has in recent years been challenged in the world of science and is not yet believed in the scientific community as it once was believed. But if it was going to be taught in schools, then I think that also the Biblical theory of creation, which is not a theory but the Biblical story of creation, should also be taught.'

He had earlier taken this attitude as Governor of California where the State Board of Education ruled in 1969 that Creation and evolution should be taught as equal, alternative theories. By 1980, Creation had become included in the public school science syllabuses of Wisconsin, Missouri, South Dakota, California and Texas and, in a number of other states, local school boards had made similar provisions. In all these instances the Creationists had achieved 'equal time' by regulation or decree. However, in March 1981 the government of Arkansas enacted the 'Balanced Treatment of Creation-Science and Evolution-Science Act'. (Interestingly, this occurred only thirteen years after that same legislature had removed from the statute book a 1928 law prohibiting the teaching of evolution.)

Out of this legislation arose a test case brought against the State of Arkansas by the American Civil Liberties Union. Among the plaintiffs were the Methodist, Catholic, Episcopal and African Methodist Episcopal bishops of Arkansas, plus representatives of the Presbyterian, South Baptist and United Methodist Churches, the American Jewish Congress and the Union of American Hebrew Congregations. Many biologists and geologists were called as expert witnesses. The defendants included the Arkansas Department of Education and several of its committees and officers. Their witnesses were mainly Creation 'Scientists'.

The trial was held in December 1981 and resulted in a clear verdict on all counts for the plaintiffs. In his decision, the judge ruled that Creation 'Science' is no more nor less than a particular religious doctrine and that a law enforcing its teaching in government schools is therefore unconstitutional. The Act has therefore been removed from the Arkansas statute book.

Scientists around the world can take some comfort from this case — for it could conceivably have gone the other way — but it can be no more than cold comfort. While it would have been very convenient for fundamentalists to have the teaching of Creation 'Science' blessed by law, absence of legislative enforcement will not reduce their continued pressure on individual school boards and parent organisations in the USA and elsewhere.

In the absence of any parliamentary debate or legislation Queensland has banned the use of MACOS and made it appear mandatory for appropriate emphasis to be given to Creationism and evolution in the state high school science syllabus. Queensland is the base for Creationism in Australia and it is interesting to quote from one of the leaders of the movement in that state:

> The Creation Science Ministry teaches both the Biblical and scientific facts of creation. The aim is to show how this teaching is related to the Gospel and basic to the truth of the Fall of man.
>
> The technique for getting our books in a library is to approach the parents and churches with advice. The principal of a school may be commended for his work in this field and our books offered for his library shelves...
>
> Especially is it pleasing to find the books in the right sections of the

libraries under Science and History instead of Religion as the atheistic Dewey system would require.

In a debate, two points of view are assumed to be of equal importance . . . Hence the rejection of the method of debate as a teaching tool in our ministry.

Since the commencement of this ministry, the aim has been to present a low profile to the community at large, and in particular the scientific community.

The Education Department funds seminars to instruct teachers in methods to teach creation . . . The impression on the educational system of Queensland has been significant and prayer has surely been the key. Through the power of prayer the majority can be influenced by a small minority . . . (Ham, 1980).

This author is more open than many of his American colleagues but his remarks were made nearly five years ago. Since then, the Queensland Creation ministry has become more subtle in its published pronouncements.

CONCOMITANTS OF CREATIONISM

Evolution is not the only thing to which fundamentalists object. We have seen that fundamentalism arose in response to a sense of betrayal of Christian beliefs and what was seen to be a decay of moral values after World War I. It is an anti-intellectual belief system which provides support for people who feel frightened or isolated and the present growth in membership and influence of fundamentalist churches may well reflect an increased sense of insecurity in the Western world.

Fundamentalists continue to oppose the teachings of the orthodox churches and, in their view, theism is as abominable as atheism. Additionally, they are vocal in upholding the sanctity of marriage, the family, male superiority, parental authority, corporal (and often capital) punishment and strong government. Divorce, abortion, homosexuality and socialism are anathema. The movement is politically conservative and in the USA it shares membership with the John Birch Society, the Moral Majority and other extreme right-wing bodies. Most fundamentalists view the USSR as the begetter of the Antichrist and many of them, as 'born-again' millenarian Christians, see Armageddon as a welcome event in the near future. As the Reverend Jerry Falwell, leader of the Moral Majority, remarked in his 1985 Oxford Union debate with Mr David Lange, true (fundamentalist) Christians need not fear an atomic holocaust for they will all be gathered up to heaven. Presumably the evolutionists will go to hell since, as the director of the Institute for Creation Research avers: 'Satan himself is the originator of the concept of evolution . . .' (Morris, 1980).

HOW TO OPPOSE THIS NONSENSE

First, it would be a great mistake to regard the Creation 'Science' movement as a fad that will fade. It is a force to be reckoned with. Every opportunity must be taken by scientists and other concerned citizens to oppose the fatuity and duplicity of the Creation 'Science' case. Indeed, *exposure* is the best weapon that can be used against a movement that seeks to exert its influence insidiously, presenting 'a low profile to the community at large, and in particular the scientific community'. Education departments and their ministers should be put under pressure to reaffirm the secular basis of Australian public education and, should they fail to do so, be taken to court for imposing a particular religious doctrine in the guise of scientific education.

Secondly, effective opposition requires continued effort on the part of scientists to present the evidences of a universe and an Earth whose ages must be measured in billions of years, and for the evolution of life over a comparable period. No science is easy and it must be recognised that many science teachers, although well based in physics and chemistry, are poorly equipped to teach the earth or life sciences. These teachers are particularly vulnerable to Creationist propaganda and will remain so unless appropriate in-service training is provided by well-qualified biologists.

Thirdly, we must continue to remind the public that the attack on evolution is only one aspect of the inroads being made by militant fundamentalism into the values which Western democracies take for granted. It is a deeply authoritarian and anti-intellectual movement that would, if given the opportunity, establish an intolerant theocratic regime. The freedom that it seeks ('the right to equal time') is the freedom to dictate.

Finally, while it may be too much to expect the orthodox churches to come to the defence of science, they should be encouraged to express their public opposition to this travesty of Christianity.

Table 1.1
Sequence of Creation events

Genesis 1:1–2:3
Day 1.   Light (day) separated from dark (night).
Day 2.   Heaven formed above Earth; waters separated into those above Earth and those on it.
Day 3.   Waters on Earth gathered into seas; dry land appears. Vegetation comes from the Earth.
Day 4.   Sun, moon, stars created.
Day 5.   Birds and all aquatic species created.
Day 6.   Reptiles, mammals and humans created.
Day 7.   Rest day.

Genesis 2:4–2:22 (no chronology)
1. No plants growing, because of absence of rain.
2. First man (Adam) formed from dust.
3. Garden of Eden planted by God.
4. Adam placed in Eden; permitted to eat all plants except 'tree of knowledge of good and evil'.
5. All wild animals and birds formed (out of soil).
6. Rib taken from Adam by God and formed into subservient woman (Eve).

Genesis 6:7–8:20 (long after Creation, confused chronology)
1. Disappointed with humans, God decides to extinguish all living organisms.
2. Changes mind, decides to save Noah and his family, together with seven pairs of every bird species, seven pairs of every 'ritually clean' species, and one pair of the remaining 'unclean' species. Noah takes one pair of each into the Ark.
3. Flood lasts ten months, rises 7–8 metres (15 cubits) above highest mountains, eliminating all living organisms except those in Ark.
4. Earth repopulated by inhabitants of Ark.

CHAPTER TWO

EVOLUTION AS SCIENCE: ONE ASPECT OF A VERY LARGE UNIVERSE

Michael Archer

'Studying science doesn't make one
a scientist any more than studying
ethics makes one honest. The
studies must be applied. Forming
and testing hypotheses is the
foundation of science, and those
who refuse to test their hypotheses
cannot be called scientists — no
matter what their credentials.
(Schadewald, 1982: 17)

INTRODUCTION

A crisis looms. While scientists have become rapt in the utterly fascinating
business of exploring the nature of the universe, of which the evolutionary
process is a part, they have as a group failed in the important responsibility
of adequately communicating this understanding to the general community. A consequence of this failure has been the rapid growth of a vocal
minority who incorrectly perceive an inevitable consequence of scientific
enquiry to be the discovery that there is no God. They fear that the study of
evolution will reveal a God-less origin of, and God-less purpose for, life.

Two vitally different matters are being confused here. The business of
science is to observe, explore and understand every possible aspect of the
material universe without regard to the existence or otherwise of supernatural phenomena. It cannot be the business of science to pass judgement
on the existence of spiritual phenomena because it is impossible to subject
such matters to scientific analysis (Wielert, 1983).

The business of religion is to understand the nature of the *non*-material or
spiritual part of the universe such as the possible meaning and purpose of
life. It is not the business of religion to determine 'on the basis of scripture'
limits for the material nature of the universe; nor is it the business of
religion to deny the existence of natural processes because these processes
appear to contradict scripture (Ferre, 1983).

To confuse science and religion will inevitably lead to pointless conflict. No human, scientist or other, can ever know that there is no God because there is simply no way that science can test such a hypothesis. Kingsley (1928) cautioned about such things, noting that failure to find 'water-babies' does not entitle one to conclude that no 'water-babies' exist. Similarly, there is no possible way to show that God could not have anticipated and determined the course of evolution such that humans appeared according to His plan. These concepts are about spiritual matters and are of an entirely different kind to the concepts of science. They in no way conflict with the concepts of science and in fact the two peacefully coexist within the minds of many religious evolutionists.

For example, Dr David Ride, a highly respected evolutionist in the Australian scientific community (and the academic supervisor of my own doctoral research), points out:

'as a scientist and a Christian, I see no threat by science to Christianity but, rather, a challenge: a challenge to theologians to make known to scientists the position of modern theological scholarship on creation, and a challenge to scientists, like myself, to express the position of science on evolution in such a way that it is clear that it addresses questions that can be examined by scientific methods, but does not deny the reality of other issues and truths about which science says nothing.' (Ride, 1985: 10).

Similarly, a Christian botanist recently asserted in the Uniting Church's journal 'A belief in God as creator, and an acceptance of evolution are in no way alternatives. I do not believe in evolution, nor in gravity. They are facts of life. I believe in God the Father Almighty, maker of heaven and earth. That is worth believing!' (Rogers, 1984: 8).

Creation 'Scientists', a vocal minority of religious fundamentalists who believe that any scientific observation which appears to contradict a literal interpretation of the King James version of the Bible is in error, epitomise the worst sort of irrationality that can result from confusion of religious and scientific concepts. They cannot or will not see the differences between the material and spiritual components of the universe. Because the Bible, which they regard as representing the infallible word of God, appears to provide statements about the nature and time of origin of the material world, they (e.g. Morris, 1974a) declare that the truth of these statements is part of an all-or-nothing package involving belief in God. In other words, they hold that to accept the concepts of evolution (which they regard as a refutation of the Biblical account of Creation) automatically involves rejection of God.

In reality it means nothing of the kind. Because the concept of evolution is one of science, it can have no bearing on the reality of a spiritual God, miracles or any other supernatural phenomena.

As well as misusing the Bible, Creation 'Scientists'' declaration that one either accepts the literal truth of every word of the Bible (as they interpret it) or one rejects God represents the worst possible sort of blackmail. The aim of this extraordinary threat is to intimidate Christians into blindly accepting Creation 'Science's' views as the only way to accept God. This kind of ideological threat would have been more at home during the thirteenth-century Inquisition than in the twentieth century. Surely people may find God in many different ways and as surely it is the right of no-one to deny to someone else acceptance of God simply because they do not like that person's understanding of science. The Creation 'Scientists'' declaration is not only illogical; it is an intolerant and un-Christian point of view.

In any case, unlike the Creation 'Scientists' who have lost much of their grip on reality because of their sworn allegiance to accepting no evidence from the world that does not support the absolute inerrancy of the word of the Bible (Ruse, 1982), most Christians find no difficulty in accepting the concepts of evolution as well as the Genesis account. Their faith, unlike that of Creation 'Scientists', does not involve determining for God the limits of His powers. They acknowledge the possibility that God may have created and guided the evolutionary process. In this perfectly sensible view, evolution may well have been God's method of Creation. As many deeply religious people have pointed out, it says nowhere in the Genesis account precisely how God created and it would seem more than just a bit presumptuous to declare that God was not allowed to create life by whatever means He chose.

How, after all, if God did use an evolutionary process to bring about His Creation could he have explained to the human transcribers of the Bible the intricacies of molecular genetics as part of this creative process, even if he had wanted to? Neither the human transcribers nor the intended audience for the Bible had any of the vocabulary or any of the essential concepts to comprehend, let alone write down, even the most elementary description of cellular biochemistry, amino acid sequences, messenger RNA, self-replicating nucleic acids, population genetics and so forth. If God's method of Creation was the same complex evolutionary process which we are just beginning to understand (all of us that is except Creation 'Scientists'), the only thing God could have said that the people of the time would have understood was, 'I created life'. The important message being conveyed was that God was the Author of the Creation; that concept humans needed to understand and this was a perfectly adequate means of conveying that understanding. The 'how' and 'when' of Creation are by comparison trivial matters and, because of the complete lack of an adequate vocabulary, were probably best explained in mythical terms. Had the descriptions been literal, they could not have been written in any language known to men for the next 3000 years (the Yahwist account including the story of Adam and Eve was probably written in the time of Solomon (Hyers, 1983)). What then would have been the purpose of providing the Bible with a literal account of God's methods of Creation if it could not have been understood for another 3000 years?

Modern discovery of the mechanisms and reality of evolution could be our first real glimpse into the mechanisms devised by God to bring about His Creation. In this view, the phenomenal complexities involved in the evolution of life would be one of the most stunning manifestations of God's power rather than, as Creation 'Scientists' conclude, a refutation of the whole concept of God.

As a demonstration that a great many deeply religious Christians perceive Creation 'Science' as something less than Christian, the following resolution was adopted by the 67th General Convention of the Episcopal Church meeting in New Orleans, 5–15 September, 1982:

> 'The terms 'Creationism' and 'Creation-Science'. . . do not refer simply to affirmation that God created the Earth and Heavens and everything in them, but specify certain methods and timing of the creative acts, and impose limits on these acts which are neither scriptural nor accepted by many Christians. The dogma of 'Creationism' and 'Creation-Science' as understood in the above contexts has been discredited by scientific and theological studies and rejected in the statements of many church leaders. 'Creationism' and 'Creation-Science' is not limited to just the origin of life, but intends to monitor public school courses, such as biology, life science, anthropology, sociology, and often also English, physics, chemistry, world history, philosophy, and social studies'.

Similarly, Pope John Paul II, in an address to the Pontifical Academy of Science in 1981, said:

> 'The Bible itself speaks to us of the origin of the universe and its make-up, not in order to provide us with a scientific treatise but in order to state the correct relationships of man with God and with the universe. Sacred scripture wishes simply to declare that the world was created by God, and in order to teach this truth it expresses itself in terms of the cosmology in use at the time of the writer. The Sacred Book likewise wishes to tell men that the world was not created as the seat of the Gods, as was taught by other cosmogonies and cosmologies, but was rather created for the service of man and the glory of God. Any other teaching about the origin and make-up of the universe is alien to the intentions of the Bible, which does not wish to teach how heaven was but how one goes to heaven'.

It is relevant as well as ironic to point out here that the Creation 'Scientists'' demands for a literal interpretation of the Bible do not even find justification in Biblical tradition. As Voorhees (1985: 8) comments:

> 'Mainline Catholics, Protestants and Jews may agree on little else, but they are agreed that scripture contains essential truths, which vastly transcend the literal text. . . There are innumerable references in both Old and New Testaments to the necessity to explain, interpret, help,

guide, teach; none of this would be necessary if the scriptures had no meaning beyond the literal words. The early church fathers, notably St Augustine, clearly rejected literalism in favour of the much deeper and more significant spiritual truths contained in scripture. In fact, no one ever considered that the Bible should be read literally until after the industrial revolution . . .'.

Presuming the Bible was divinely inspired, it was clearly God's intention that the Bible be interpreted rather than accepted blindly.

Creation 'Scientists' seem determined to ignore the centuries of wisdom accumulated by Biblical scholars in the same way that they dismiss centuries of accumulated science. Not surprisingly, within the ranks of the Biblical fundamentalists there is growing, even now, a move to reinstate geocentricism in astronomy (the view that the whole universe actually moves around the Earth every 24 hours) because this is a view that can be defended if Genesis is taken literally (Schadewald, 1985).

The absolute faith placed by Creation 'Scientists' in the literal word of the Bible (and then only just the King James translation) represents bibliolatry, the worship of a book. To use this bibliolatry as the sole basis for deciding who is and who is not allowed to find God, what is and what is not the correct way to perceive the structure of the material as well as the spiritual universe, and what could or could not have been decisions made by God to effect His Creation, seems, to the majority of Christians and scientists alike, a fatuous, intolerant and quite possibly idolatrous view.

Creation 'Scientists' vehemently attack the concept of evolution as if it were the brain-child of Satan (Morris, 1967, 1974a, 1977) rather than the deductions of modern science. They *know* they are right so have no need to consider data obtained from the material world other than to attack it if it appears to threaten their beliefs. As the militant Creationist Billy Sunday said, 'When the word of God says one thing and scholarship says another, scholarship can go to hell.' (McLoughlin, 1955).

As part of their strategy they have begun a campaign to have the teaching of a literal interpretation of the Genesis account of Creation included in the science syllabuses of schools and universities (see Strahan, this volume).

But, Creation 'Science' as it is practised is, by its own definition, simply not science (Gish, 1974b, 1985; Morris, 1967, 1974a, 1977; Whitcomb and Morris, 1961a). It is religion. For this very simple reason, it should not be taught in science classes (Edwords, 1983; Godfrey, 1983; Kitcher, 1982; Montagu, 1984). If it is to be taught at all, its place is in a course on comparative religion alongside the hundreds of other different cultural mythologies that have just as much right as Creation 'Science' to claim perfect but non-scientific understanding of the nature of Creation and the universe.

Not unexpectedly, Creation 'Scientists' do not agree. They have already managed to insinuate the teaching of Creation 'Science' into the Queensland education syllabus alongside the concepts of evolution (Thulborn, 1985b). Further, at the insistence of the Queensland Minister for Education, Mr L. Powell, the Queensland Department of Education has arranged special seminars with the 'Creation Science Foundation' of Queensland (*Creation Science Prayer News* for March 1985). These seminars for Queensland teachers are apparently provided so that they might effectively teach this religion in science classes.

What was once thought of by Australian scientists as an American anti-intellectual anachronism has suddenly sprung up smack in the middle of our own educational system.

Being aware of the possible confusion between methods of studying the spiritual and material aspects of the universe facing science teachers and students, I have made a four-pronged attack on the problem. This chapter continues with a brief explanation of the modern understanding of how evolution occurs. Later chapters give an overview of the evidence in support of the concept and Creation 'Scientists'' objections to it.

WHAT IS EVOLUTION?

The term 'evolution', as Mayr (1982) points out, has had many meanings. In its loosest sense, it is used to describe any change in any thing. In a much stricter biological sense, to modern geneticists it is 'any change in genetic makeup in populations of organisms'. This definition of biological evolution, while strictly correct, does not include any of the particular processes or products now accepted as being implicit in the concept. Accordingly, evolution can be defined more fully as: 'The origin of life from prebiotic substances and the subsequent differentiation through time of all species from pre-existing species, this ongoing process being the result of changes produced by natural selection and/or mutation in the genetic makeup of populations'.

Although Charles Darwin's basic concept of evolution, as set out in *On the Origin of Species by Means of Natural Selection or the Preservation of Favoured Races in the Struggle for Life* (Darwin, 1859), still survives among contemporary evolutionists, it has undergone considerable modification. Many additional concepts have resulted from modern evolutionary studies (e.g. Curtis, 1983; Luria *et al.*, 1981; Mayr, 1982; Nei and Koehn, 1983; Stebbins and Ayala, 1985). For example, we now understand the basic genetic mechanisms of evolution that were unknown to Darwin and as a consequence we realise that there are many more ways in which populations change and species arise and survive than he visualised in 1859 (Gould, 1977; Stebbins and Ayala, 1985; White, 1978). Biologists who use this wider 'synthetic' array of concepts to solve evolutionary problems have been referred to as 'neo-Darwinists'.

The modern idea of how evolution occurs involves basically five concepts.

Overproduction of young

Evolutionary change can be a consequence (but not an inevitable one) that follows from overproduction of young by any type of organism; that is, birth of more young in each generation than can survive to reproduce. For example, a wild cat may have six kittens despite the fact that there may not be enough resources in the environment for that many additional cats. As a result, some of the kittens will die before they reach adulthood. The number of excess young produced is commonly inversely related to the size of the organism. For example, a virus may produce millions of 'young' in excess while an elephant may only produce ten young in excess during its lifetime. However, much (if not most) evolutionary change occurs without differential mortality. Other factors such as genetic drift (see below) have to be considered.

Variation and its heritability

Normally, offspring in a new generation will not be identical — some of the wild cats will be smaller than others, some will have longer ears or different colour patterns, some may meow half an octave higher than the others and so forth. Darwin knew that much of this variation between young was heritable (i.e. capable of being passed on to the next generation), although he did not understand the mechanism. Today we know that this variation and its heritability has its basis in the genetic material of all organisms and, more specifically, in the DNA molecules of the chromosomes (see any modern biology textbook such as Curtis (1983) or Luria *et al.* (1981)). Further, it is now clear that new variation can arise by non-directed mutations in the genetic material of the parental organisms and that such new variations can be either useful, harmful or of no consequence to the survival of the young in which it occurs.

Natural Selection

In addition to deaths due solely to chance, there will be an inevitable and natural culling in each generation of the individuals that compete unsuccessfully for the limited resources in their environment. Insofar as the variation acted upon has a genetic basis, this selective culling will remove from the population at least some of the genetic instructions for the competitively unsuccessful structure, function or behaviour. The sum of factors responsible for this culling is what Darwin called natural selection.

Change through time

The constant action of natural selection on each new generation can result in a progressive shift through time of the genetically determined attributes of the population. Alternatively or coincidentally, the genetic makeup of a population can also change due to the accumulation of random mutations that have no selective value. This type of random change is called genetic drift. No matter what the cause, all shifts in genetic makeup, in the strictest sense of the definition, constitute evolutionary change.

Speciation

If a portion of the species becomes isolated (e.g. the forest habitat of our wild cats might be broken into two isolated and smaller forests by a fire), the two portions may (given enough time) diverge sufficiently in structure and/or behaviour so that even if individuals from the two isolated groups were reunited they would no longer be willing or able to interbreed successfully. By that stage, they would have become different species. A modern definition of a biological species is (Mayr, 1982: 273): 'a reproductive community of populations (reproductively isolated from others) that occupies a specific niche in nature'. The emphasis here is on reproductive isolation.

An Australian example of how geographic isolation can lead to reproductive isolation and speciation can be seen in the recent history of whipbirds (genus *Psophodes*) (Keast, 1958). Two species are of interest — the eastern whipbird (*P. olivaceus*) which occurs in dense thickets near the wet forests of the east coast and the western whipbird (*P. nigrogularis*) which inhabits the drier coastal sandhills and mallee scrub of southwestern Western Australia and parts of southern South Australia and northwestern Victoria (Figure 2.1). The parental stock from which the two species developed lived in appropriate habitats at a time when a continuous forest belt linked the east coast of Australia to southwestern Western Australia. Distribution of the parental stock is shown in black in Figure 2.1(a). Climatic deterioration during the Pleistocene caused changes in vegetation and loss of appropriate habitats, leading to isolation of two populations — one along the east coast and one in the southwestern corner of the continent (Figure 2.1b and c). Isolation of the western stock led to speciation as the population became adapted to drier conditions and began to inhabit mallee scrub (hatched area in Figure 2.1d). Climatic fluctuation, reflected in vegetation change and hence available habitats, allowed the western isolate to spread back across the Nullarbor (Figure 2.1e). Further climatic change caused loss of habitat in the Bight region and led to the development of two isolated populations of the western whipbird — one in Western Australia, the other in the Mallee of Victoria (Figure 2.1f). Lack of appropriate habitats between the Mallee and the forest belt in Victoria has prevented the ranges of the two species from developing any overlap.

Speciation can occur rapidly if natural selection pressures are intense. However, speciation can also occur without the action of natural selection. There can be a steady accumulation of neutral mutations in an isolated portion of a population of a species (genetic drift) which can result in reproductive isolation. There can be a sudden mutation, restricting an individual's capacity to interbreed successfully with any individual within the population other than one carrying the same mutation (or with itself, as in some plants).

Mechanisms of reproductive isolation in closely related species are very varied. They include: behavioural mechanisms — such as alterations in courtship rituals; chemical mechanisms — such as changes in sex-attracting hormones; and physical mechanisms — for example, changes in reproductive organs. These mechanisms inhibit mating between individuals. Often, however, individuals of reproductively isolated species willingly mate. Then, genetic mechanisms — such as chromosomal mutations — exist which prevent cell division or make the developing zygote inviable.

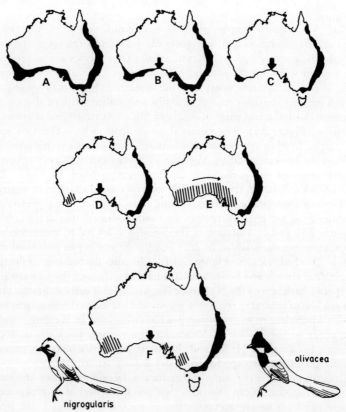

Figure 2.1.   Speciation in *Psophodes* (Australian whipbirds). See text for a full explanation. Broad arrow indicates loss of habitat, and the narrow one the eastward spread of western population.

Until complete reproductive isolation occurs some seemingly different species may be able to form hybrids. In some cases the hybrids are infertile — for example, offspring of swamp wallabies (*Wallabia bicolor*) and agile wallabies (*Macropus agilis*). This hybrid infertility means that the two species will remain isolated. At the other extreme, hybrids between grizzly bears (*Ursus arctos*) and polar bears (*U. maritima*), produced in captivity, are fertile — suggesting that the two forms of bear are a single species. In nature the two types are geographically isolated and unlikely to mate. They may be in the process of developing reproductive barriers.

To look at this whole question of evolution in another way, we could ask 'what conditions would have to be met before evolution could *not* occur?' There are four basic conditions: (1) mutation would have to be impossible because mutations can change the genetic makeup of a population; (2) natural selection would have to be impossible because this changes the frequency of alternative genes in a population; (3) a population's size would have to be infinite, otherwise random genetic drift could lead to loss of a particular gene and hence a change in genetic makeup of the population; and (4) there would have to be no migration of individuals with slightly different genetic makeups between particular populations of the same species because this would alter the gene frequencies of the populations.

In nature, there is not a single population of any organism that could fulfil these four conditions. The consequence is simply that evolution is an inevitable attribute of life. Some of the evidence for this conclusion will be considered below.

The study of evolution now incorporates concepts drawn from many other areas of science including biochemistry, biophysics, palaeontology, morphology, ecology, behaviour, population genetics and so on. It is difficult (and probably pointless) to try to define the range of subjects woven into the study of evolution. While its conceptual framework is broad, the study of evolution (as will become evident below) is a very healthy science.

THE CONCEPTS OF EVOLUTION AS SCIENCE: CAN THEY BE TESTED?

The methods by which science attempts to achieve understanding about the nature of the universe commonly involve the following steps: (1) observations are made; (2) a falsifiable hypothesis is developed that would best explain these observations; (3) this hypothesis is tested by the gathering of more observations in an effort to check the hypothesis (in practice, the originator of the hypothesis often tries to find supportive evidence while colleagues try to find contradictory evidence!); and (4) depending on the outcome of the test, either the original hypothesis is rejected in favour of another that would more adequately explain (i.e. predict) the new observations as well as the old, or more observations are gathered in a continuing effort to check the original hypothesis. Although some good ideas in science have arisen by shrewd guess work and often pig-headedness in holding on to hypotheses despite evidence to the contrary, most of the steady progress of science has involved these four steps.

This self-checking aspect of science is a very vital part of the whole process. When Creation 'Scientists' suggest, as they often do, that scientists are involved in some sort of conspiracy to protect conventional scientific hypotheses from being falsified, they betray either their extreme naivety or deliberate refusal to understand what actually happens. In fact, I have never heard of a practising scientist, evolutionist or other, who at heart does not delight in discovering a rotten spot in the core of his research field. Such a discovery, if successfully defended in publications in refereed research journals, frequently leads to research money to explore further related aspects of the field. A perusal of any issue of *Nature*, for example, will reveal that almost as much space is devoted to the refutation of previous views (in evolutionary biology as well as other fields of science) as is devoted to the presentation of entirely new concepts.

That is not to say that the individuals who propose new hypotheses will as enthusiastically seek evidence to falsify them. It is, perhaps regrettably, human nature for the author of a hypothesis to defend it vigorously against all but the most devastating disproofs. But you can be absolutely certain that the colleagues of these authors will leap at the opportunity to falsify the hypothesis and will waste no time in publishing contradictory evidence if they can find it. In a sense, this is a form of natural selection and the survival of the fittest among hypotheses. The system of scientific procedure as a whole thrives on continuous self-checks and ruthless self-correction. For these reasons, any attempt at deceit or conspiracy in real science will eventually prove to be futile.

Because one of the most important precepts of the scientific method is that no hypothesis is ever beyond the reach of additional testing and the possibility of falsification (Popper, 1959), there can never be an infallible 'Bible' of scientific knowledge. In the long run, the only useful ideas in science are those that can be falsified. The potential for refutation is vital in science for without it we would have no way of judging the relative worth of ideas that compete to explain some aspect of the universe.

In terms of its credentials as science, evolution is in large part testable (Popper, 1978; Ruse, 1982; Williams, 1982, and see below). All of the theoretical concepts of evolution as set out above have been repeatedly tested and have so far survived. For example, we can experimentally test whether or not a hypothetical type of selection pressure occurs, either by setting up a controlled experiment in a laboratory or by going out into the wild to determine the outcome of predictions that follow from the evolutionary hypothesis.

At the same time, it should also be noted that there are some evolutionary ideas that cannot be tested experimentally. For example, the hypothesis that evolutionary processes observable today were the sole basis for species transmutations seen in the fossil record cannot, for obvious reasons, be tested experimentally. However, of all hypotheses about the origin of the natural world, the evolutionary model has been shown many times over to be a much better predictor of the known fossil record than any of its competitors (such as Creationism; see Ruse, 1982, and the examples given below). For example, mid-Ordovician jawless fish from the Northern Territory, middle Triassic mammal-like reptiles, Late Triassic triconodont mammals, the Jurassic *Archaeopteryx lithographica* and the Pliocene *Australopithecus afarensis*, all fossil forms that are structurally and temporally intermediate between other previously known types of organisms, were predictable types of organisms to find within the framework of the evolutionary model. They are not predicted by Creationist models and, in fact, in some cases falsify the Creationist models.

There are of course areas of conflict amongst theorists of evolution, such as the debates about punctuated equilibria versus gradual evolutionary change (see the papers in Chapter 5 of Smith, 1982b and the papers in volume 11(1) of *Paleobiology*, 1985). These are not, as is sometimes suggested in Creationist literature, about the *reality* of evolution itself; rather they are about possible differences in the *rates* of evolutionary change. Similarly, debates presently being waged about phylogenetic systematics versus other types of phylogenetic studies (see, for example, the chapter on systematics in Archer and Clayton, 1984) are not about the reality of evolution but about the best way to reconstruct evolutionary pathways.

Conflicts among evolutionary theorists, although often misinterpreted by Creationists as evidence of basic flaws in the evolutionary model, are a part of any healthy science. Without this sort of constant testing and refining, hypotheses can all too readily slide out of the realm of science and into that of dogma. And, if growth in the amount of research involving evolutionary topics is any indication, evolution science is in no danger of such a slide.

FURTHER READING

Cherfas, J. (ed.) (1982) *Darwin up to date*. IPC Magazines Ltd (as 'A New Scientist Guide'), London.

Dobzhansky, T.D., Ayala, F.J., Stebbins, G.L. & Valentine, W.H. (1977) *Evolution*. W.H. Freeman, San Francisco. 572 pp.

Forey, P.L. (1981) *The evolving biosphere*. British Museum (Natural History), London & Cambridge University Press, New York. 311 pp.

Gould, S.J. (1982) Is a new and general theory of evolution emerging? *In* Smith, J.M. (ed.) *Evolution now: a century after Darwin*. *Nature* and Macmillan Press Ltd, London: 129–145 (This article was originally published in *Paleobiology* 6: 119–130.)

Lewontin, R.C. (1974) *The genetic basis of evolutionary change*. Columbia University Press, New York.

Margulis, L. (1981) *Symbiosis in cell evolution*. W.H. Freeman, San Francisco.

Mayr, E. & Provine, W.B. (eds) (1980) *The evolutionary synthesis*. Harvard University Press, Cambridge, Mass. 487 pp.

Nei, M. & Koehn, R.K. (1983) *Evolution of genes and proteins*. Sinauer Associates, Sunderland, Mass. 331 pp.

Patterson, C. (1978) *Evolution*. British Museum (Natural History), London & the University of Queensland, Brisbane.

Smith, J.M. (ed.) (1982) *Evolution now: a century after Darwin*. Nature and Macmillan, London.

Stansfield, W.D. (1977) *The science of evolution*. Macmillan, New York. 614 pp.

CHAPTER THREE

THE REALITY OF ORGANIC EVOLUTION: EVIDENCE FROM THE LIVING

Michael Archer

What I am trying to show is how evolutionists have been able to establish that the kind of theoretical picture contained in population genetics actually finds a response in nature — the theory is about things that really occur. (Ruse, 1982: 103)

INTRODUCTION

When Charles Darwin was gathering observations prior to his publication in 1859 of *On the Origin of Species by Means of Natural Selection, or the Preservation of Favoured Races in the Struggle for Life*, the vast body of data available to him came from the living world. Although he was profoundly influenced towards his notion of evolutionary descent by encounters with the fossil vertebrates of Europe, South America and Australia (because the fossil creatures he encountered from each of these continents resembled most the living forms that succeeded them in the same region), not enough was then known about the fossil record to enable him to know about transitional series or significant 'links' between otherwise different sorts of creatures.

As a result, Darwin developed his theory of evolution almost entirely on the basis of studies of living organisms. Even in this area he was severely handicapped because he had inadequate knowledge of the mechanisms of heredity. Hence he did not have an appropriate basis for understanding the way in which the variations he observed originated, nor for that matter how these variations, once acquired, were passed on to future generations. This understanding did not really come to be part of the modern understanding of evolution until the early part of the twentieth century.

It is appropriate to ask how, in the more than 100 years that have elapsed since Darwin published *On the Origin of Species*, his theory of evolution has fared. Has subsequent scientific research supported the concept of evolution or falsified predictions implicit in his model? We could answer this question with a simple 'Yes, it has been supported; no, it has not been falsified', an entirely correct response (Ruse, 1982), but, for anyone interested in the actual evidence, not a very satisfactory one. So, let's consider some of the actual evidence in support of evolution. We will do so in two parts: that which comes from the living world (this chapter) and that which comes from the fossil record (Chapter 6).

In a later chapter we will also consider some of the current objections to the evolutionary model. Most of the objections have involved attempts to falsify one or more of the concepts of evolution and/or to challenge the observations interpreted by others as supporting the model.

EVIDENCE IN SUPPORT OF EVOLUTION FROM THE MODERN WORLD

There are thousands of books and research articles whose principal conclusions have been that studies of the structure and function of living organisms support the theory of evolution or at least some aspect of the model. It is beyond the scope of this chapter to review all of these. Instead, the interested reader should turn to some of the excellent summaries of this literature available (see Further Reading).

Here we will consider a few examples of the sort of data, experimental as well as observational, that have been gathered from living organisms.

The steady production of new genetic variation

Variation is the resource upon which the process of natural selection must act in order for evolution to occur. Darwin recognised the existence of variation but did not understand its nature or origin. It is now clear that the basis for genetic variation is the sequence of adenine, cytosine, guanine and thymine base-pairs in the DNA (or, in some viruses, RNA) contained in the chromosomes. Differences in the sequence of base-pairs can ultimately result in differences in the proteins manufactured in the cells and these in turn can result in differences in an organism's morphology, function and behaviour (Dobzhansky *et al.*, 1977; Lewontin, 1974; Patterson, 1978; Stansfield, 1977).

Most genetic-based differences in character combinations that occur between generations are the result of cross-overs between the DNA-containing segments of the two parental chromosomes or recombination of segments on different chromosomes. These cross-overs can occur while the cells of the organism are reproducing or dividing in preparation for the production of gametes. While this results in most of the variation upon which natural selection acts during each generation, it is not the way in which *new* genetic variation develops over time.

New genetic variation develops by means of mutations that occur during replication processes of cells. These 'mistakes' in replication are essentially random events and *not* the immediate result of natural selection. Mutations may be small spontaneous changes, known as point mutations, in the coding sequence of sections of the DNA molecules within the chromosomes. This sort of mutation occurs while the DNA molecule is replicating. Spontaneous large-scale chromosomal mutations can occur when the nucleus of the cell is about to divide and can involve the complete loss of part of a chromosome, end-to-end inversion of a chromosome section or the addition of whole extra sets of chromosomes.

Among the best-known point mutations in humans are those which have occurred in the particular parts of the DNA that code for production of haemoglobin molecules. At least 250 different varieties (mutant forms) of haemoglobin are known, including that for sickle-cell anaemia, and it is now clear that approximately 1 in every 2000 people has mutant haemoglobin. The rate at which mutations occur in the parts of the DNA molecules that code for haemoglobins in a wide variety of organisms is at least 1 in 100 000 individuals per generation. The rate may be higher because not all of the mutations are detectable. The rate can also vary depending on the amount of radiation or mutagenic chemicals to which a person is exposed. Point mutations in general are rarely harmful and may be one way in which life accumulated, over billions of years, much of its original variety of genetic information.

Although this point mutation rate might seem too low to be significant, when you take into consideration the total amount of DNA replicating and the number of cells in our bodies continually undergoing replication, it is probable that we each accumulate during our lives somewhere between 1000 and 10 000 point mutations of this sort. Of these mutations, however, only the ones that occur in the reproductive cells can become new variation available for transmission to the next generation.

In contrast to point mutations which result in new genetic variation, chromosomal mutation involves rearrangements of existing genetic variation. Chromosomal mutations involve substantial rearrangement of chromosome material such as inversion of chromosome segments, crossovers between inappropriate chromosomes, loss or superabundance of genetic information and so on. Mutations resulting in the loss of chromosome segments are commonly lethal, making it impossible for the remaining DNA to produce a viable individual. Other types of chromosomal mutations, such as duplications, may be beneficial because they enable the individual that possesses them to pass more genetic variation into the next generation (Lewontin, 1974; Patterson, 1978).

To summarise, modern research has demonstrated not only that existing genetic information may be reshuffled by natural means into new combinations, without adding or subtracting information, but that mutations are constantly producing entirely *new* genetic information as grist for the mill of natural selection. It is further clear that these mutations, depending on their nature, may be harmful (e.g. some types of chromosomal mutation), helpful or of no apparent significance to the organism's well-being (e.g. many of the haemoglobin point mutations noted above).

Natural selection, the motive force of evolution

The reproductive potential of an organism normally exceeds the number of its descendants that survive. The factors that determine which offspring survive to pass their genetic information into the next generation constitute what is called natural selection. The individuals that contribute the most genetic information to the next generation are said to exhibit the highest 'fitness'.

Many examples of natural selection have been demonstrated in the laboratory and in the wild and it is clearly a general phenomenon (see Further Reading). Despite this fact, there is no point in trying to test a *general* concept of natural selection — apart from its application in population genetics where mathematical tests are constantly carried out on predictions about the fitness of genes (see the discussion in Kitcher, 1982) — because it is not a principle or general theory. It is a simple description of reality. It is a term for a kaleidoscope of processes or events (such as predation by marsupial quolls, competition between carnivorous ghost bats and owls or the effect on two species of competing seed-eating ants of a sudden flush of nutrient-rich *Acacia* seeds) whose reality and effects are specific to the particular situation being considered.

While the concept of natural selection is clearly an important aspect of the evolutionary model, the reality of evolution as a whole does not depend on a demonstration of the reality of every instance of natural selection. On the other hand, it *is* possible and appropriate to test hypotheses about the reality of natural selection in any particular case.

What is important here is to consider whether there are examples, experimental or otherwise, of selective processes that have produced changes in the genetic composition of organisms — i.e. evolutionary changes. That is the first matter to consider. Whether or not these selective processes also improved the 'fitness' of the organisms they may have changed is a different, although clearly related, matter.

Industrial melanism

Consider the well-known example of industrial melanism in the British peppered moth, *Biston betularia*. Few high school biology texts fail to mention this study yet few students (and almost no Creationists) seem to understand what it is that this example demonstrates.

In all the insect collections of the eighteenth and early part of the nineteenth centuries in England, individuals of this moth exhibit pale wings flecked with black markings which is the variety of the moth known as *typica*. Then, in 1848, a black (melanic) individual was caught near

Manchester. By the 1880s, the black individuals, which were named *carbonaria*, outnumbered the pale individuals and by 1895, 98–99% of all individuals caught around Manchester were black. The same melanistic trend in *Biston betularia* occurred in other areas and in over 200 other species (Lees, 1981).

The appearance and rapid increase in numbers of *carbonaria* coincided with the rapid growth of industrialisation, associated with which was a steady increase in the amount of coal burned. In areas where coal burning increased, consequent pollution destroyed the pale coloured lichen that previously grew on the dark tree trunks and on which the moths rested by day.

Because the peppered moth is hunted by birds, it seemed reasonable to hypothesise that predators would select (by eating) more of the conspicuous light moths resting on the dark (non-lichen-covered) trunks than they would the dark moths.

In experiments to test this selective predation hypothesis, marked dark and light moths were released into both pollution-affected and non-pollution-affected woodlands. Birds were subsequently observed in these study areas searching for moths around the trunks of trees. When surviving marked moths were collected in light traps and counted there was a greater percentage of pale survivors in the non-polluted area with lighter-coloured tree trunks and a greater percentage of dark survivors in the polluted area. Calculations of the intensity of natural selection through bird predation show that this pressure to tip the balance in favour of either light-coloured or melanic moths was very high. Similar natural selection pressures during the second half of the 1800s would have been sufficient to explain the rapid spread of the dominant *B. b. carbonaria* mutation once it had appeared.

It is now also clear that the dark form, *B. b. carbonaria*, is a mutant in which a single point mutation has produced a dominant condition; both homozygotes and heterozygotes with the mutation are dark. Presumably this mutation had occurred before but it was not until the advent of industrial pollution that it became advantageous to moths carrying it.

An interesting postscript to this example of natural selection in action is the fact that since efforts have been made through legislation to reduce industrial pollution, the proportion of pale *Biston betularia* in previously heavily polluted areas has increased.

Clearly, environmental pressures, through natural selection, can effect rapid shifts in the genotype of a population. In this case the spontaneously produced variation that proved to be advantageous to the species' survival was a genetic mutation. This is evolution in action, under observation. What it is not (nor was it ever claimed to be despite what one may find in Creationist literature) is an example of the evolution of a new species.

Effects of drought on Darwin's medium ground finch

The year 1977 was a major drought year for Daphne Major, an islet in the Galapagos Islands. It received only 24 mm of rain, an amount that was less than a fifth of the yearly average. Between 1975 and 1978, studies were carried out on the biology of the seed-eating Darwin's medium ground finch (*Geospiza fortis*) and more than 1500 birds were colour-banded and measured (Boag and Grant, 1981).

During 1977, the species did not breed at all on the island and by the end of the year had suffered an 85% decline in numbers (only one of the 388 nestlings marked in 1976 had survived). Analysis of the mean size of adults before and after the drought revealed a marked shift in mean size with most survivors being larger individuals with larger bills.

A similar analysis of seeds over this period revealed that more larger and harder seeds survived the drought than smaller, softer seeds.

Boag and Grant demonstrated that the increase in size and hardness of seeds was highly correlated with the increase in size of the surviving finches over the same period, the best explanatory hypothesis being that larger birds with larger beaks are more capable of coping with larger and harder seeds. Because approximately 76% of the variation in the measured morphological features had been demonstrated to be heritable, the significant increase in body and beak size constituted a rapid evolutionary change brought about by intense natural selection.

Resistance to insecticides

As another demonstration of the reality of evolutionary change brought about by natural selection, we can consider a matter of considerable economic importance to us: insect resistance to insecticides. Wood and Bishop (1981) have reviewed studies of the evolutionary changes in the genotypes of insects that have occurred as a result of the use of particular insecticides.

Pest species of insects almost invariably develop strains that are resistant to particular insecticides. Similarly, bacteria frequently develop resistance to particular antibiotics which is why it often becomes very difficult to treat some human diseases.

Figure 3.1 summarises the development of resistance to particular insecticides used to control the housefly (*Musca domestica*) between 1945 and 1972 in Denmark. What is immediately obvious is the short space of time within which this insect developed resistant strains. The rate at which advantageous mutations appeared in these populations is very high and, because the resistance is heritable, the insecticides soon became ineffectual. Once a mutation conferring resistance has arisen, further use of the insecticide provides an intense selection pressure in favour of flies carrying that mutation.

Genetic studies of insecticide resistance have demonstrated another interesting point. Resistance to a particular insecticide in different populations of the same species does not always have the same genetic basis. This is again a measure of the capacity of intense selection pressure (in this case manmade) to effect an evolutionary change in the genotype of an organism.

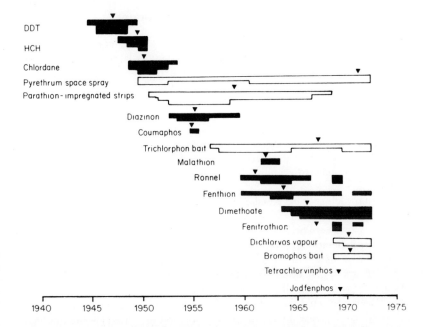

Figure 3.1. Development of resistance to various insecticides in Danish populations of the housefly *Musca domestica*. Duration of use of each pesticide is shown by length of the bar; thickness of the bar indicates extent of use. Solid bars indicate what were residual sprays; open bars represent other methods of insecticide application. Appearance of resistance to a pesticide is indicated by a small black triangle. (After Wood & Bishop, 1981.)

FROM GENETIC CHANGE TO REPRODUCTIVE ISOLATION: THE EVOLUTION OF SPECIES

It is one thing to demonstrate that there is ample evidence for evolutionary change within a species. It is quite another to demonstrate that an existing species is being transformed into a new species. Yet this is what the evolutionary model predicts. Is there any evidence for it?

Quite simply, the answer is 'yes!'. Recall that the definition of a biological species is: 'a reproductive group of populations (reproductively isolated from others) that occupies a specific niche in nature'. The key concept is reproductive isolation. Processes that lead to reproductive isolation can result in speciation. Evidence that reproductive barriers can develop comes not only from laboratory-based experiments but also from observations of populations in the wild.

Speciation processes observed in the laboratory

Many of the laboratory studies of speciation follow genetic change in flies of the genus *Drosophila*. They are easy to breed under laboratory conditions, have a quick turn-around time between generations and produce a large number of young. We will look at three examples.

Drosophila pseudoobscura and *D. persimilis* are species of vinegar fly, individuals of which will, under experimental conditions, mate with members of the opposite species 4–6% of the time. Such matings result in fertile females but sterile males. A series of experiments, detailed in Kessler (1966), was conducted to see if, by applying a selection pressure, the tendency of flies to mate with members of the other species could be modified.

Within each species, Kessler established a line of flies selected for a low tendency to mate with the other species and a line of flies selected for a high tendency to do so. Selection in each line of each species was carried out for eighteen generations. Eventually, flies of both species, selected for a low tendency to mate with the other species, were allowed to interbreed. Flies of both species, selected for a high tendency to mate across the species boundary, were also allowed to interbreed.

Lineages selected for a low tendency to mate with the other species showed an increase in reproductive isolation — cross-species matings dropped from 4–6% to 0–2%. Lineages selected for a high tendency to mate with the other species showed an increase in the tendency: from 4–6% to as high as 21% in some cases. The tendency of flies to mate across species boundaries had obviously been modified.

A similar laboratory study was carried out to determine if isolated populations of *Drosophila melanogaster*, exposed for long periods to different natural selection pressures, would become reproductively isolated. For five years, two isolated populations were raised in environments that differed in temperature and humidity. At the end of the experiment it was found that, while individuals within each population bred successfully, barriers to reproduction had begun to develop between the two populations.

Another study involving *Drosophila pseudoobscura* was carried out by Powell (1978). He too wanted to test whether mimicking a natural selection pressure could result in the development of barriers to reproduction between populations. He was testing the hypothesis that isolation followed by expansion of a very small part of a population (i.e. one with a restricted subset of the genetic variation present in the whole of the parental population) may be a stage in the development of reproductive isolation.

After only three cycles of isolation and expansion of a randomly chosen small part of the population, third-cycle descendants of the first isolates showed a definite disinclination to breed with members of the parental population from which the isolate had been taken.

Powell's experiment demonstrated that if a population is reduced to very few individuals (such as could occur if a small group of individuals colonised an island), that population may eventually become sufficiently different from the parental population for reproductive isolation to occur. This can happen because the gene frequencies (and hence variation) of the 'founding' individuals will be only a subset of those present in the parental population. Subsequent expansion of a population based on such a subset will result in a significantly different descendant population. Because this experiment demonstrated that one attribute of this difference may be unwillingness to interbreed in the individuals of the two populations, it also demonstrated how easily populations may be ushered to the brink of a speciation event.

Speciation processes at work in the wild

Sticking with flies of the genus *Drosophila*, a study by Prakash (1972) revealed that what has been demonstrated experimentally in the laboratory can also be seen to occur in the wild. During extensive surveys of Colombia in 1955 and 1956, no *D. pseudoobscura* were found. The nearest population occurred about 2400 km away in Guatemala. Then, in 1960, individuals of this species began to turn up in insect traps in Bogotá, Colombia. They had either dispersed to Colombia naturally or been accidentally introduced by humans. However they arrived, they soon became one of the most common species of vinegar fly in Colombia.

Prakash examined the capacity of individuals from the new Bogotan populations to reproduce successfully with those from elsewhere. He found that although Bogotan males mated successfully with females from other areas, Bogotan females produced totally sterile males when crossed with males from other areas. In the short span of time during which the Colombian populations had been isolated a new form had developed, already reproductively different from other populations of vinegar flies.

A similar study was carried out by Bush (1969, 1975a,b) involving the apple maggot fly *Rhagoletis pomonella*. When first recognised, this North American fly's host plant was the native North American hawthorn (*Crataegus*). The mean time of emergence from the soil of the pupating larvae coincides closely with the period when hawthorn fruits ripen between early September and early November. Male flies are territorial; as soon as they locate a ripe hawthorn fruit they defend it until a female arrives. Females will not approach a male unless he is defending a ripe fruit. After mating, she lays her eggs in the fruit he has found. After the maggots have fed and are ready to pupate, they leave the fruit and enter the soil. The emergence of adult flies and their mating behaviour is tightly tied to the time of maturation of the hawthorn fruit. If it were not, the flies would not find a host plant in which to deposit their young.

In 1865, about a century after apple trees were introduced to North America, the first infestation of apples by *Rhagoletis pomonella* was noted. Now, apples in North America begin to ripen in late July (approximately four to five weeks before the hawthorn fruits start to mature) and continue until late September. So, in order for apples to become a host species, individual flies within the population must mature before the time for breeding successfully in ripe hawthorn fruit. Before introduction of the apple, early-maturing variants would normally have failed to find a suitable fruit in which to breed and died without leaving offspring. With the appearance in the *same* region of an earlier-maturing fruit, the once less 'fit' flies turned out to be fortuitously right on time for the new resource.

Studies of these flies carried out by Bush reveal that in the one area there are now two coexisting forms: the original hawthorn 'race' and a new apple 'race'. The two 'races' have quite different mean times of maturation. Some populations of the two 'races' now even differ in relative body size, ovipositor size and the number of bristles behind the eyes. In other words, within little more than a century, one ancestral population has been observed to diverge into two different forms with different times of reproduction. To what extent temporal isolation of the different forms extends to other reproductive isolating mechanisms has yet to be determined.

The last type of evidence we will consider here to test whether or not speciation processes are at work in the wild (or for that matter in the laboratory) involves completed speciation events produced by spontaneous chromosomal mutations. The most common of these events involves production of what are called polyploids (Gibby, 1981; White, 1978).

Polyploids have three or more *haploid* sets of chromosomes in each nucleus. They can result from spontaneous mistakes in the segregation of chromosomes during cell division. If polyploidy occurs in the sex cells, it can be an instant passport to reproductive isolation from all other non-polyploid individuals in the population. Because it depends on accidents to chromosomes during cell division, unlike point mutations, it is a speciation mechanism that does *not* involve natural selection.

Tetraploids, with twice the normal number of chromosomes in the nucleus, are among the most common type of polyploids and are known to occur in a wide variety of organisms, including humans. In one study of human embryos that had died before birth about 1% were found to be tetraploids. In another study of naturally-aborted human embryos, 4% were found to be triploids (i.e. the result of fusion of a normal haploid gamete with an unreduced diploid gamete; Gibby, 1981). While there is no indication here that triploid or tetraploid humans are viable (though polyploidy may not have been the cause of death), there is a clear indication that this sort of chromosomal mutation in 'higher' vertebrates is a uncommon event. In any case, it is clear that many worms and some crustaceans, insects, molluscs, fish and amphibians are successful polyploids (Gibby, 1981; White, 1978).

It is also evident from modern studies that this sort of speciation event has occurred in many groups of plants. It has been estimated, for example, that 47% of the species of flowering plants arose through polyploidy. An impressive catalogue of known polyploids is given in White (1978). We will consider just one example: *Triticum aestivum* (a tetraploid wheat which we eat every day).

The first species of wheat known to be cultivated by humans was a 14-chromosome diploid species *Triticum monococcum* (einkorn wheat) that is still cultivated in Asia Minor. Genetic studies have shown that another wheat, *T. turgidum*, is a tetraploid 28-chromosome species that probably arose from chromosome doubling in a natural hybrid between einkorn wheat and possibly *T. speltoides*, a 14-chromosome grass species that grows in the same fields. *Triticum aestivum*, which is modern bread wheat, subsequently arose as a 42-chromosome tetraploid hybrid between *T. turgidum* and *T. tauschii*, a 14-chromosome diploid species also found in the Near East. Exactly when *T. aestivum* first arose in the wild is uncertain, but both *T. turgidum* and *T. aestivum* have been 'recreated' in the laboratory as chromosomal mutations during hybridisation experiments with the species presumed to have given rise to them in the wild (White, 1978).

Known polyploids include many other naturally-occurring plants that have subsequently been 're-evolved' in the laboratory during hybridisation experiments between related species taken from the wild. New fertile species have arisen for the first time in cultivation (e.g. *Primula kewensis*, a tetraploid that arose spontaneously in a botanic garden; White, 1978). New fertile polyploids (e.g. *Raphanobrassica*, a polyploid hybrid between a radish and a cabbage) have even been developed experimentally from hybrids between species of different genera!

Clearly both speciation processes and completed speciation events have been observed in the wild and duplicated as well as initiated in the laboratory. This is unambiguous evidence in support of the evolutionary model and for the hypothesis that evolutionary change should be able to lead to the development of new reproductively isolated species.

GEOLOGICAL HISTORY AS SUPPORT FOR EVOLUTIONARY HYPOTHESES

Before passing on to the fossil record in a search for tangible evidence of previous evolutionary events, we should briefly consider two quite different experimental approaches, using living organisms, that test the reality of historical evolutionary events.

An example of the first approach is Carson's (1976) study of Hawaiian species of *Drosophila*. Previous studies involving many groups of organisms had suggested that the degree of difference in the structure of enzymes as revealed by electrophoresis could be a reasonable measure of the time elapsed since those organisms shared a common ancestor (Ayala, 1976). This was hypothesised to be so because many studies had demonstrated that point mutations were constantly occurring in the DNA coding for each enzyme. Carson's study compared electrophoretic differences in enzymes among Hawaiian vinegar flies with the known time of origin for each of the Hawaiian Islands on which they occur.

The ages of the volcanoes that make up the five largest Hawaiian Islands are well known (Macdonald and Abbott, 1970). The oldest island (Kauai) is about 5.7 million years old and the youngest (Hawaii) about 700 000 years old. The five islands, which form a linear chain with the oldest in the northwest and the youngest in the southeast, formed sequentially as the Pacific crustal plate passed over a 'hot spot' in the Earth's mantle.

Using electrophoretic techniques, Carson examined the enzymes of eight closely related but morphologically distinct Hawaiian species of *Drosophila*, each of which is restricted to one or at the most two of the islands.

He found that the two species which occur on Hawaii differ very little in enzyme structure, as might be expected if they had had the least time to differentiate. When the enzyme differences between all of the species were plotted against the known ages of the islands on which they occur there was a close linear relationship; the greatest enzyme differences occur between species that had been separated for the longest time, the least in those recently separated. An apparent exception was suspected from other (cytological) evidence of having originated on one island and dispersed to another.

The evidence for evolutionary divergence amongst these flies, based on modern electrophoretic studies, is independently supported by the geological age of the separate islands on which they occur.

MOLECULAR CONFIRMATION OF EVOLUTIONARY HYPOTHESES

One way to test evolutionary hypotheses is to determine the patterns of relationship between species suggested by independent character systems not involved in the development of the original hypothesis. One of the many character systems used in tests of evolutionary hypotheses is the sequence of nucleotides coding for the proteins of living animals.

Let's consider an example. Suppose that on the basis of analyses of the gross morphology of eleven mammals (using features such as tooth shape, skull structure, ankle bone morphology and so on), two of the species are hypothesised to have had a common ancestor which they shared with none of the other nine. We should be able to test this hypothesis of relatedness by comparing the nucleotide sequences coding for one of the proteins shared by all eleven species. Since most (if not all) changes in nucleotide sequence occur by mutation, all reproductively isolated species should steadily diverge from one another as a function of time. The degree of divergence will then be a function of time elapsed since the organisms shared a common ancestor.

This being so, we would expect to find that species which on the basis of studies of gross morphology are suspected of having shared a common ancestor unique to themselves would also share unique nucleotide sequences.

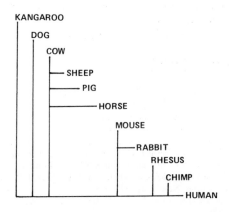

Figure 3.2. Relationship between eleven mammalian species inferred from nucleotide sequences coding for five different proteins. Note the basic confirmation of pre-existing evolutionary hypotheses (based on morphology, serology and other studies) of the particularly close relationship between chimp and human; between all of the ungulates (cow, sheep, pig and horse) and between all placentals (all forms shown except the kangaroo). (After Penny, Foulds and Hendy, 1982.)

This type of test was carried out by Penny, Foulds and Hendy (1982) in an examination of the evidence in support of evolutionary hypotheses about the relationships between eleven mammal species (Figure 3.2). Rather than being content to examine differences in sequences of nucleotides coding for one protein of these mammals, they examined sequences coding for five different proteins. The results from each of the different proteins were strikingly similar and taken together were strong support for the basic evolutionary hypotheses. As they concluded: 'The general conclusions from the present work are that (1) it is possible to make falsifiable predictions from the hypothesis that species have been linked in the past by an evolutionary trees and (2) there is strong support from these five sequences for the theory of evolution.'

SUMMARY

So, to recapitulate, there are both laboratory and field data to support the hypothesis that new species can and do evolve from pre-existing species. In the examples considered, mechanisms demonstrated to be capable of leading to reproductive isolation include natural selection and chromosomal mutation. Theoretically, once it has been demonstrated that evolution occurs and that new species can in fact arise from pre-existing ones, there should be no difficulty in perceiving that by these same means all of the species on Earth could have evolved from a common ancestor. But here again, we do not have to accept the reasonableness of this hypothesis solely on the basis of evidence from the modern world; we can test it using the evidence of the fossil record. That evidence we will review in Chapter 6.

FURTHER READING

Cherfas, J. (ed.) (1982) *Darwin up to date*. IPC Magazines Ltd (as '*A New Scientist Guide*'), London.

Forey, P.L. (1981) *The evolving biosphere*. British Museum (Natural History), London & Cambridge University Press, Cambridge. 311 pp.

Godfrey, L.R. (ed.) (1983) *Scientists confront Creationism*. W.W. Norton., New York.

Hedrick, P.W. (1983) *Genetics of populations*. Science Books International, Boston. 629 pp.

Hutchinson, P. (1974) *Evolution explained*. David & Charles, London.

Jukes, T.H. (1983) Molecular evidence for evolution. *In* Godfrey, L.R. (ed.) *Scientists confront Creationism*. W.W. Norton, New York: 117–138.

Mayr, E. (1982) *The growth of biological thought*. The Belknap Press of Harvard University Press, Cambridge, Mass. 974 pp.

Mayr, E. & Provine, W.B. (eds) (1980) *The evolutionary synthesis*. Harvard University Press, Cambridge, Mass. 487 pp.

Nei, M. & Koehn, R.K. (1983) *Evolution of genes and proteins*. Sinauer Associates, Sunderland, Mass. 331 pp.

Smith, J.M. (ed.) (1982) *Evolution now: a century after Darwin*. Nature and Macmillan, London.

CHAPTER FOUR

THE UNIVERSE UNFOLDS

Ronald Brown

A PRETTY ORDINARY PLANET

The science of astronomy is providing us with an increasingly detailed knowledge of the structure of the universe. In the period of Kepler, Galileo and Newton we started to understand the nature of the Earth in relation to the sun and something about the sun's nearest neighbours — the planets of the solar system. By the nineteenth century it became widely accepted that the stars were considerably further away than the planets but it was not until the twentieth century that astronomers started to realise that the sun was part of a great collection of stars now estimated to number between 10^{11} and 10^{12} which form an object called a spiral galaxy — known to us as the Milky Way.

Neither the sun and its planetary system nor the galaxy to which they belong is unique. From the study of spectra of stars we know that the sun is *a very ordinary and common type of star* — many stars are about the same size, luminosity and temperature. Technically it is classed as a G2 star based on the nature of the spectrum it emits. In a G2 star, such as the sun, the spectrum contains conspicuous ionised calcium lines and weak hydrogen lines; other ionised and neutral metal lines are present as well. The nature of the spectrum is a function of the surface temperature. A G2 has a surface temperature of $5000-6000°K$ — in the sun actually $5800°K$ — and appears yellowish white.

There is a continuing debate in astronomical circles as to whether planetary systems such as ours are commonplace or exceedingly rare, as was once thought to be the case. We have no direct evidence of other planets outside this solar system because we cannot see them. They are illuminated by reflected stellar light and are therefore too dim to be observed against the background of stellar glare. However, we do have indirect evidence for their existence. This comes from the displacement in the path of a star caused when a massive object, such as an invisible planet, passes close by. These stellar wobbles are detectable only for those few stars that are relatively close to the sun. On the basis of more indirect evidence, such as the rotational periods of stars, it has been suggested that there may be more than 10^{10} stars with planetary systems in *our* galaxy. Perhaps when the first space telescope becomes available we will get first-hand observations of other planetary systems. Until then we must accept that their existence is the subject of scientific argument.

THE MILKY WAY AND OTHER GALAXIES

If the Milky Way has any unique characteristic it is that we live in it. Otherwise we know from photographic evidence and from extensive studies of the heavens that the Milky Way is one of many spiral galaxies. There are other kinds of galaxies too, and in the last few decades the study of extra-galactic astronomy, i.e. the universe outside the Milky Way, has revealed a great number. In all, the total number of galaxies has been estimated to be at least 10^{11}. The galaxies are clumped together in clusters and it is starting to appear that these clusters are strung together in a rather open network throughout the universe. But on this gigantic scale it is difficult to come to highly detailed pictures of the overall arrangement of the matter.

It has been found that the more distant galaxies are all moving away from the Milky Way. This has been detected by means of the famous 'red shift' observed first in sound waves by an Austrian scientist, Christian Doppler, and later applied to light by Armand Fizeau, a French scientist. The 'red shift' is caused by movement. It was detected by observing the spectral lines in moving and stationary objects. When an object emitting light is moving away from an observer, the position of the spectral lines, which occur at specific wavelengths in a stationary object, are moved (shifted) towards the red end of the spectrum. The observation of the 'red shift' in light from distant galaxies has led to the concept of an expanding universe in which matter on the large scale is steadily moving apart. The density of matter must therefore be diminishing with time, an important problem to be confronted by cosmological theories.

Because of the finite velocity of light, the light that reaches telescopes on Earth today originated in distant galaxies a very long time ago. This means that we are able to observe objects at a much earlier stage in the evolution of the universe. If there is a noticeable evolution of galaxies with time these distant objects could represent primitive stages of young galaxies. If there are similarities between these objects and others closer to the Milky Way they perhaps would reveal to us the way in which the Milky Way galaxy originally formed. This has raised questions as to whether distant objects are similar to the objects that are closer to the Milky Way, a particularly important question if one is interested in the way in which galaxies might evolve with time.

It has also been found that some of the distant objects have no counterparts closer to the Milky Way. The objects that have attracted perhaps the most attention are called quasars (quasi-stellar objects). Present thinking is that they may represent the nuclei of distant galaxies, these nuclei undergoing very violent events that make them abnormally bright. It may be that this is a phase through which some kinds of young galaxies pass.

However, my role is not so much to describe our present knowledge of the universe but rather to consider our current views on the way in which the universe has evolved.

COSMOLOGICAL THEORIES

Cosmological theories attempt to account for the nature of the universe on the largest scale, not only as it exists today but as it evolves with time. Although there has been a variety of cosmologies over the history of mankind, very few of them have been based on fundamental physics. This is not surprising since it was not until this century and the development of the general theory of relativity by Einstein that a suitable theoretical basis was available. Einstein's theory can be thought of as a first fundamental theory of the nature of gravity and it is gravitational forces that overwhelm all others when one is considering the largest scale phenomena of all, the nature of the entire universe.

THE EXPANDING UNIVERSE

It has been possible to describe the expanding universe by use of some suitable solutions of Einstein's equations. A consequence of these solutions, too complex to examine here, is that the age of the universe can be determined very simply as the reciprocal of the constant (known as the Hubble constant) relating the expansion velocity of a galaxy and its distance from Earth. The constant is named after an American scientist who observed, earlier this century, that galaxies further from Earth moved with greater speed. For every additional million light years distance from Earth the velocity of the galaxy increased by 170 km s^{-1}. Thus Hubble's constant (H) is in kilometres per second per million light years. Distance and speed are also related by the familiar equation, distance/speed = time. Thus time = $1/H$, i.e. assuming that the universe is expanding at a uniform rate the reciprocal of the Hubble constant represents the time when expansion began, i.e. the age of the universe. Because of problems in determining the very large distances that are involved with galaxies and quasars, there is still strong scientific debate about the value of the Hubble constant. Astronomers now believe the value for H lies between 50 and 100 km per sec per megaparsec (Mpc) in modern units; where 1 Mpc is one million parsecs (pc) and 1 pc = 3.26 light years = 3.09×10^{13} km. Thus, astronomers are uncertain of the value of H over a range covered by a factor of two. This means that we are uncertain of the age of the universe by the same factor.

Now, this uncertainty may be an encouragement to non-scientists (particularly Creationists attempting to justify their estimates of the age of the universe) to declare that scientists really don't know the age of this universe. For them to do so would demonstrate a singularly uneducated view of the nature of science. All of us recognise that when any measurement is made there will be an uncertainty about the value obtained. For the age of the universe uncertainty spans in the range of 10×10^9 to 20×10^9 years. *Most importantly for deliberations about the notions of Creationists, despite uncertainty, the age determined astronomically is still many orders of magnitude different from an age of 6000 years.*

This method of estimating the age of the universe depends on the assumption that the observed red shifts of spectral lines are the result of the Doppler effect. There have been several attempts to devise alternative interpretations of the red shift which do not require it to be related to a velocity of departure of the source from the observer. Suggestions have been made that the velocity of light might change with time or that photons steadily lose their energy as they move throughout space. These suggestions, to the best of my knowledge, come to grief. For them to be valid entails serious contradictions in some other area of physics. There is at present no credible alternative to contemplate other than the concept of an expanding universe.

THE AGE OF THE STARS

When early estimates of the age of the universe were made using the first value of the Hubble constant there was some unease among astronomers. This was because these early estimates gave an age for the entire universe of 1.8×10^9 years, i.e. somewhat younger than the age of the older stars, calculated to be about 10^{11} years. Estimates of the age of the stars came from a general understanding of the processes which happen within stars, arising from studies of the stars and the way they evolve. The theory of stellar evolution incorporates the corresponding time scale on which stars go from their initial formation, by condensation from more tenuous gas, to their expiry, either relatively gently through a red giant stage to a white dwarf or more violently through flaring up as novae or supernovae. The calculations of age rely heavily and almost entirely on the rates at which nuclear processes take place, i.e. the way in which protons collide with each other or other particles and are converted into helium nuclei and other larger nuclei. The rates of these processes have been carefully measured in a number of physics laboratories throughout the world, leading to very substantial quantitative agreement among scientists. Thus the theory of stellar evolution is based on very thoroughly developed physics. Some of the finer details are still being worked over but they will not alter the general picture outlined above which has remained a stable part of astrophysics for some years.

Accordingly there is no real dispute amongst astronomers about typical lifetimes of the stars of various kinds. For example, the life expectancy of a star like the sun is approximately 10^{10} years. It has been evolving for about 5×10^9 years and has just another 5×10^9 years to go. Other stars and groups of stars have been found in the universe that are estimated to be older than the sun; and in the objects known as globular clusters (groups of perhaps 10 000 or more stars) that are in orbit around the Milky Way galaxy the stars are known to be very old, in some instances perhaps slightly more than 10^{10} years. It is gratifying to see that refinements in establishing distances of distant galaxies have led to revised values of the Hubble constant that give an age for the universe which is in substantial agreement with the age of the oldest known globular clusters of stars.

NUCLEAR CHRONOMETERS

A further confirmation of our current view of the age of the universe is provided by another area of nuclear physics that also has been thoroughly researched, one for which recently Fowler at California Institute of Technology shared the Nobel Prize for Physics. From observations of various isotopes that undergo radioactive decay with very long half-lives a number of so-called nuclear chronometers have been devised. From the observed amounts of various isotopes in stars and other astronomical objects there is again good evidence that these objects have ages of 10×10^9 years or perhaps rather more.

The nuclear chronometers and the ages of stars in globular clusters have been derived without any reference to things such as the velocity of light or the wavelength of photons undergoing Doppler shifts. One cannot therefore reconcile these observations with the very much younger age of the universe derived from hypotheses relating to light. Instead one has to produce some other hypothesis that results in a changing time scale for events concerned with nuclear processes.

THE PRODUCTION OF ELEMENTS

Let me now turn to still further aspects of the evolution of the universe. The broad scenario is that in the early moments of the universe there were mostly rather exotic particles but these rapidly evolved into protons and neutrons, the neutrons mostly being rapidly converted in their turn into helium nuclei. So the very early universe was composed overwhelmingly of just *two* of the chemical elements that we know today, i.e. hydrogen and helium.

Before more complicated atomic nuclei could be generated the hydrogen and helium of the universe had to evolve through the processes of formation of clusters of galaxies to the stage where sufficiently dense clumps of gas emerged and condensed into the first generation of stars.

Such stars, being composed virtually entirely of hydrogen and helium, did not, in the language of the astronomers, contain any metals (to an astronomer a metal is any element other than hydrogen and helium).

Elements such as carbon, nitrogen and oxygen and a number of other elements are produced in substantial amounts only towards the end of the life cycle of a star when it is much cooler than at its formation. So one has to wait a very long time for the cosmic appearance of these elements. In the case of the sun it will take about 10^{10} years to get to the stage of production of *metals* but in the case of more massive stars the life cycle is shorter, perhaps as short as 10^7 years in some instances. The metals comprising our planet Earth (and you and me) were made in the nuclear furnaces of an early generation of large short-lived stars, our solar system having been formed from their debris. The point I am making is that even in the most enthusiastic estimates, time intervals of at least 10^7 years, and more realistically considerably longer than this, are needed to explain the formation of all heavier elements. The stuff that you and I and all other living creatures on Earth are made of and indeed much of the stuff of the Earth itself could not exist if the universe was as young as 6000 years! I have seen no scientifically credible hypothesis that could explain the production of the elements on shorter time scales.

THE FORMATION OF PLANETARY SYSTEMS

Throughout the history of astronomy there has been a steady flow of proposals advanced as to how our sun and planets originally formed. During at least the eighteenth, nineteenth and the current centuries most of these proposals made use of the principles of physics known at the time. Nearly all of them were found to contain serious problems in relation to some aspect of the physics of the processes described. From these we have now distilled only one or two ideas that seem to be consistent with physics as a whole.

I will not attempt to describe them in detail but will focus on the concept that is most widely adopted by astronomers as consistent with all observations so far. It is that great clouds of gas and dust, called dark nebulae, gradually form in various places throughout the Milky Way and other spiral galaxies of the universe. The clouds consist of heavy metals formed in the stellar nuclear furnaces of an earlier generation of stars and spewed out into space when these stars expired. When such clouds get massive enough or cold enough they collapse, either under their own influence or through the effect of some external influence such as a shock wave (as might be generated, for example, by the explosion of a nearby supernova). The process of collapse leads to the formation of new stars but it is inevitably a rather untidy process leaving other matter outside the newly formed stars. This other matter ends up as the planets, asteroids, comets and interplanetary dust; all the objects that are now studied by planetary astronomers. The presence of dust mixed with the gas is essential for the process to take place. The dust is composed of metals which, as explained above, cannot be formed until the stars have virtually completed their life cycle. Consequently the first generation of stars had to form from a pure gaseous mixture of hydrogen and helium. They could not have been accompanied by planets, comets and the rest.

A number of interesting chemical processes occur throughout the life of these gigantic clouds of gas and dust, a subject to which my own research group has devoted quite a bit of its time in recent years. We have now reached the point where we have theoretical models that appear to account very well for radio-astronomical observations on these clouds. These chemical models imply that the lifetime of such objects is of the order of millions of years. The deduction of this type of age is based largely on chemical reaction rates in the laboratory and so is independent of other astronomical observation of ages of other objects.

The detailed dynamics of the formation of planetary systems is still being developed by scientists and so we cannot make very many statements with great confidence. However, all attempts to provide a theory of the formation of the solar system require a very substantial time scale for its development. Moveover, the process of condensation of the planet Earth into a habitable planet is described in terms of a time scale involving millions of years and probably many, many millions of years. Again there are uncertainty ranges in these figures but they do not extend down to such astronomically trivial time scales as a few thousand years as Creationists would have us believe. Again these projections do not rely on the values to be adopted for one or two physical constants like the velocity of light; rather they depend on standard principles of dynamics and heat transfer in relation to condensation of liquids and solids from gaseous material.

SUMMARY

The account of the unfolding of the universe is such a gigantic scientific undertaking that inevitably the overall story contains parts that have not been worked out in the finest numerical and mathematical detail. The story that has been devised, however, is now sufficiently self-consistent that it is hard to doubt the kind of time scales and the overall sweep of the evolutionary story. The word 'evolution' used here has very little to do with the word evolution used in areas of biological science. There is no concept of 'survival of the fittest' in talking about production of galaxies and stars and planets, rather it is a picture of the inevitability of such processes occurring in the fullness of time. The future will no doubt see still further refinements of the cosmological theories but it is hard to believe that they will head in a direction that makes them look anything remotely like the story contained in the first chapter of the Bible.

FURTHER READING

Asimov, I. (1967) *The universe*. Allen Lane. The Penguin Press, London. 285 pp.
Zeilik, M. (1982) *Astronomy: the evolving universe*. 3rd edition. Harper & Row, New York. 623 pp.

CHAPTER FIVE

TESTIMONY OF THE ROCKS
OR
GEOLOGY VERSUS THE 'FLOOD'

Alex Ritchie

We know on the authority of Moses
that longer ago than six thousand
years,
the world did not exist.
Martin Luther

Some drill and bore
The solid earth
And from the strata there
Extract a register, by which we learn
That he who made it
And revealed its date to Moses,
Was mistaken in its age!
William Cowper ('The Task', 1785)

INTRODUCTION

Two hundred years after James Hutton first proposed his revolutionary ideas on the Theory of the Earth, it seems a little bizarre that geologists should find it necessary to defend the concept that the Earth is immensely old. Yet, the sciences involved in dating rocks are coming under assault from a non-scientific source — fundamentalist Creationists.

The issue is simple. It is a choice between radically different interpretations of the origin and nature of the universe and Earth and their development through geological time.

The Creationist cause has been presented in books, pamphlets, films and tapes. Many of their statements are regularly quoted in Creationist publications and can therefore be accepted at face value. They show clearly the motivation and philosophy behind 'scientific' Creationism — a concerted effort to construct a 'natural history' compatible with a literal reading of the Bible.

Fundamentalist Christians of various denominations accept the Biblical account of Genesis as the word of God and literally true. They unquestioningly accept a young-Earth model of Creation, together with its corollary, Flood Geology. These concepts are diametrically opposed to the scientific interpretation.

The creationist version is as follows:

God created the universe — matter, energy and life — in six 24-hour days. All living and extinct kinds of animals and plants were created at this time (see Chapter 1). Creation took place between 6000 and 10 000 years ago. A widely quoted date for Creation is Bishop Ussher's estimate of 4004 BC. For the fascinating story behind the Biblical chronologies of Ussher and his contemporaries see Brice (1982).

Some 1600 years after the Creation, God caused a Flood which destroyed all animals and plants except those preserved on one boat — Noah's Ark. The Noachian Flood not only tore down most pre-existing rocks but also redistributed them around the globe in the positions we find them today. Many animals and plants which perished during the Flood were preserved as fossils. After the Flood, the survivors emerged from the Ark and repopulated the bare Earth. The Flood episode lasted about one year and seventeen days.

By contrast, scientists now estimate that the universe originated between ten thousand million and twenty thousand million years ago (see Chapter 4), and that the solar system formed much later, between 4.5 and 5 thousand million years ago. The Earth, moon and meteorites consolidated around 4.5–4.6 thousand million years ago. The oldest rocks known on Earth have been dated at around 3.8 thousand million years, and the earliest evidence of life comes from rocks around 3.5 thousand million years old.

Deciphering the history of the Earth, its age and its complex and changing structure is one of the greatest feats of the human mind. We only began to comprehend the magnitude of geologic time less than 200 years ago. We have been able to measure it for about 80 years.

A sense of geologic time is one of the most important things for a geologist to convey to the non-geologist. Most people think in terms of five human generations; two behind, an emphasis on the one to which they belong, and two ahead. When it comes to envisaging the span covered by historic time, far less geologic time, the human mind has problems. We can measure it, but are not able to comprehend what John McPhee (1981) called 'deep time'. For historic time we rely on written records or artefacts to understand our history. When we come to examine the history of the Earth we face a great hurdle. For almost all of it, no human was around to witness and document it for posterity. We have to go directly to the 'book of the rocks' and, painstakingly, learn to read it.

To a geologist 'a million years is a short time — the shortest worth messing with for most problems...if you free yourself from the conventional reaction to a quantity like a million years, you free yourself from the boundaries of human time. Then in a way you do not live at all, but in another way you live for ever.' (McPhee, 1981: 129).

One of the first people to free us from the strait-jacket of Biblical interpretations with a time span of a few thousand years was James Hutton of Edinburgh.

Until Hutton's time (and for many years afterwards) igneous rocks such as basalt and granite were widely interpreted as being precipitated from water. Hutton not only recognised their igneous nature but explained how they originated from magma rising from the depths of the Earth before cooling and crystallising. He identified several sites in Scotland where basalts or granites had clearly penetrated pre-existing rocks, thus overturning earlier ideas that igneous rocks were older (or 'primary') parts of the crust.

In modern geological terms, the essence of the Huttonian Theory can be expressed as follows:

1. Processes seen today are of the same kind as those which operated in the past.
2. Basalt and granite are igneous rocks.
3. Erosion constantly modifies surface features of the Earth.
4. There is a cycle of rock change brought about by weathering, transport, deposition, folding, uplift, igneous and metamorphic activity.
5. The time available to effect these changes is enormous.

Hutton envisaged this cycle of erosion, deposition, consolidation and uplift as repeated indefinitely. His theory was supported by observations in the field, although many of the critical rock exposures were discovered after the theory was conceived.

One of Hutton's sites, near Siccar Point in southeastern Scotland, shows where older sedimentary rocks have been tilted almost vertical and deeply eroded before being covered by later, horizontal sediments — an angular unconformity, providing direct evidence of two Huttonian cycles of Earth history, one atop the other.

Hutton's younger colleague and interpreter, John Playfair, was impressed when he realised the full implications of the unconformity. 'What clearer evidence could we have of the different formation of these rocks, and of the long interval which separated their formation, had we actually seen them emerging from the bosom of the deep? . . . The mind seemed to grow giddy by looking so far back into the abyss of time.'

Hutton's ideas were more clearly expressed and popularised by Playfair (1802) in *Illustrations of the Huttonian Theory of the Earth*. 'Amid all the revolutions of the Globe, the economy of nature has been uniform and her laws are the only things which have resisted the general movement. The rivers, and the rocks, the seas and the continents have been changed in all their parts; but the laws which direct these changes, and the rules to which they are subject, have remained the same.'

These principles were further developed by Charles Lyell as the Theory of Uniformitarianism, based on the idea that 'the present is the key to the past'. The views of Hutton, Playfair and Lyell conflicted with the widely held Diluvial or Catastrophic Theory (see Hallam (1983) for a full discussion of the controversy). Modern geological theory follows the uniformitarian approach but does not take it to some of the extremes maintained by Lyell.

The scientific approach assumes that:

1. Natural laws are constant and uniform in time and space.
2. Unless there is clear evidence to the contrary, processes now operating should be invoked to explain events of the past.
3. Geological change is slow, gradual and steady.

This does not rule out the possibility of catastrophic events such as major earthquakes, volcanic eruptions, climatic changes, floods or impacts of extraterrestrial bodies.

THE GEOLOGICAL RECORD — SCIENTIFIC VERSION

The basic principles of stratigraphy are as follows:

1. Describing the *initial conditions* of sedimentary deposition.
2. *Superposition* — in any stratigraphic sequence the youngest strata are on top.
3. *Original horizontality* — strata are deposited horizontally or nearly so.
4. *Lateral continuity* — strata originally continue in all directions until they either terminate against the boundaries of the depositional basin or they thin to zero.

In the early nineteenth century, techniques were developed to interpret the rock sequences of various parts of Europe. William Smith recognised that many distinctive sedimentary formations also contained characteristic fossil remains, absent from the rocks above and below them. Using these *index fossils* to trace many formations of different ages across England, Smith produced the first geological map of the whole country in 1815. In France, Baron Cuvier and others recognised that such fossils provided overwhelming evidence for periodic mass *extinctions*. The recognition of an orderly succession of fossil faunas in widely separated places led to enormous advances in intra- and intercontinental stratigraphic correlation.

The basic tools dividing geologic time into manageable, recognisable units in a logical, testable, worldwide scheme were almost complete. The latter part of geologic time, the section for which we have a good fossil record, has been subdivided into a detailed, widely accepted hierarchy of more than 150 internationally recognised eons, eras, periods, epochs and ages. Some of the stratigraphic units probably represent less than one million years and many are applicable on a worldwide basis. These stratigraphic units are almost entirely based on the observed succession of fossil faunas and their correlation throughout the world.

The standard geologic scale for the time since the beginning of the Cambrian has been in use since the early nineteenth century (see Table 5.1). What should be immediately obvious from a study of Table 5.1 is that *all but two* of the divisions were recognised and defined prior to 1859 when Charles Darwin published his seminal work *On the Origin of Species*. Virtually all of the scientists who proposed them were not only good geologists trying to come to terms with, and make sense of Earth history, but they were, almost to a man, acknowledged Creationists. Indeed some, like Sedgwick, were clergymen. All of them came eventually, sometimes after years of struggle with their deep-seated religious beliefs, to the conclusion that the geological evidence against a single, universal Noachian Flood was overwhelming.

Sedgwick, in his Presidential Address to the Geological Society of London in 1831, said:

'Having been myself a believer, and, to the best of my power, a propagator of what I now regard as philosophical heresy...I think it right, as one of my last acts before I quit this Chair, thus publicly to read my recantation.

There is, I think, one great negative conclusion now incontestably established — that the vast masses of diluvial gravel, scattered almost over the surface of the earth, do not belong to one violent and transitory period...'

Hugh Miller, an experienced geologist who was a Creationist, could not accept a young Earth and single worldwide Flood. He put it forcefully in *The Testimony of the Rocks* (1857: 422): 'Nor can I doubt that [the Earth's] history throughout the long geologic ages...will be found in an equal degree more worthy of its Divine Author than that which would huddle the whole into a few literal days, and convert the incalculably ancient universe into a hastily run-up erection of yesterday...'

The idea peddled today by fundamentalist Creationists that the geologic time scale is an invention of evolutionists is erroneous and misleading and needs to be recognised as such. The scheme in use today, virtually unaltered since the 1850s, was developed by *Creationist* geologists. It remains in use because it is soundly based on careful observation. It is testable (or falsifiable). The idea that any serious geologist can work without accepting, or referring to, the standard geologic column is ludicrous.

THE GEOLOGICAL RECORD — CREATIONIST VERSION

Fundamentalist Creationists maintain that *all* Cambrian and post-Cambrian rocks (sedimentary, igneous and metamorphic) were laid down during the

TABLE 5.1
The standard geological scale, the scientists who defined the units and when they did so

Era	Period	Epoch	Events
Cenozoic	Quaternary	Recent	
	Morlot, 1854	Pleistocene	First humans
	Tertiary	Pliocene	
	Brongniart, 1810	Lyell, 1833	
		Miocene	
		Lyell, 1833	
		Oligocene	
		Beyrich, 1854	
		Eocene	
		Lyell, 1833	
		Paleocene	Beginning of
		Schimper, 1874	age of mammals
Mesozoic	Cretaceous		Last ruling
Phillips, 1841	d'Halloy, 1822		reptiles
	Jurassic		First birds
	Brongniart, 1829		
	Triassic		First mammals,
	Alberti, 1834		dinosaurs
Paleozoic	Permian		Last trilobites
Phillips, 1841	Murchison, 1841		
	Carboniferous		First reptiles
	Conybeare and		
	Phillips, 1822		
	Devonian		First amphibians,
	Sedgwick and		
	Murchison, 1839		last graptolites
	Silurian		First land plants,
	Murchison, 1839		coral reefs
	Ordovician		First vertebrates
	Lapworth, 1879		
	Cambrian		First animals with
	Sedgwick, 1835		hard parts

Noachian Flood. They have to *refer* to the geological column but they do not *accept* it, as made clear by Gish (1973: 42–43):

'In order to evaluate evolution as an interpretive model to explain origins, therefore, and to compare the predictive value of this model to that of the creation model, the assumptions of evolutionary geologists concerning the duration of geological ages and the validity of their assumptions concerning the geological column must be used along with the model. Therefore, in the succeeding pages of this book we will write as though the Cambrian, Ordovician, Silurian and other sedimentary deposits were actually laid down during the time spans generally assumed by evolutionists, and that the arrangement of the geological column in the form of successive geological periods as accepted by evolutionary geologists is correct.

We wish to emphasize again that we do not accept these assumptions...'

There is no attempt to argue a case, to provide an alternative interpretation of the geological column, or to provide supporting evidence for such a sweeping statement — just a bald assertion that the whole geological column is a figment of the imagination of 'evolutionary geologists'.

Gish (1973: 40–41) was equally specific on the interpretation of the fossil record:

'It is believed that most of the important geological formations of the earth can be explained as having been formed as the result of the worldwide Noachian Flood described in Genesis, along with attendant vast earth movements, volcanic action, dramatic changes in climatic conditions, and other catastrophic events. The fossil record, rather than being a record of transformation, is a record of mass destruction, death, and burial by water and its contained sediments.'

Gish (1973: 45) laid great stress on the apparent absence of Precambrian (in Creationist terms, pre-Flood) fossils: 'The oldest rocks in which indisputable fossils are found are those of the Cambrian Period....
What do we find in rocks older than the Cambrian? *Not a single, indisputable, multicellular fossil has ever been found in Precambrian rocks!...*'
This statement, often repeated in Creationist texts, is refuted by evidence, available since the 1950s and 1960s, of abundant, well-preserved fossil remains of many different types of complex, soft-bodied organisms in the Late Precambrian rocks of the Flinders Range in South Australia — the Ediacara Fauna. These discoveries have been followed by similar discoveries in Late Precambrian rocks of several other continents. For a pre-Gish popular account see Glaessner (1961) and for a comprehensive, up-to-date, worldwide review of Precambrian fossils see Glaessner (1984). Even older fossils, the remains of simple algae and organisms resembling bacteria, are now known from various parts of the world — Australia, Canada, southern Africa. We now have a patchy, but in some places, remarkably detailed record of a succession of life forms of increasing complexity through some 80% of Precambrian time.

The relationship of the Biblical Flood to the geological record and column was also spelled out by Whitcomb and Morris (1961b) and Morris (1963).

Morris (1963: 72–74):

'The flood, then must mark a great discontinuity in the ordinary geological and hydrological processes of the earth. Any geological deposits which may have existed before the flood must have been eroded, re-worked and re-deposited in some new location and sequence, such that it would not be possible to deduce now, by analogy with present processes, the geological history of the antediluvian earth. Pre-flood geography must have been completely revised, with the new continents formed by great uplifts after the flood, as well as new ocean basins for the retention of the mass of flood waters covering the earth. It follows as plainly as anything could be that any system of historical geology which ignores the fact and effects of the universal flood must be largely fallacious. Geologic dating methods which assume continuity of rates of the associated geologic processes, as all such methods necessarily do, can only be valid for a period of time extending, at most, back to the flood.

. . . Out of all the apparent chaos of destruction that must have been caused by the flood it is apparent that the final deposits which resulted from it would have assumed a certain statistical order!

For example, there would naturally be a tendency for those sediments and organisms which occupied the lowest elevations before the flood to be buried deepest by the flood. Thus simple marine organisms and marine sediments would tend to be buried deepest, then fishes and more complex marine creatures, then reptiles and amphibians, then mammals, and finally, man.

Another factor controlling the order of deposition of the sediments and the organisms contained in them would be that of the relation between the specific gravity and the hydrodynamic drag. . . . Thus, there would be a tendency for organisms of high density and simple structure to settle out most rapidly and, therefore, to be buried deepest. This factor of hydrodynamic selectivity is often highly efficient. . . .

The simpler, less mobile, smaller creatures would thus be caught and trapped first, whereas higher animals and especially man would often be able to retreat to the very highest points in the region before being inundated. This, too, would mean that most men and higher animals would never be buried at all in the sediments, but would float on the surface of the waters until decomposed or destroyed by scavenger fish.

. . . Thus, the fossil deposits and sediments in which they are contained can very logically be viewed as an actual historical record preserved in tablets of stone, of the terrible events of the year of the great flood.'

Morris (1963: 58–59):

'. . . The Biblical framework, therefore, requires that we categorically reject the fossil record as a record of the history of the 'development' of life on earth.

They [the fossils] must have been laid down *after* the introduction of the present order of things into the universe and deposited under the action of the present physical laws which now control the behaviour of nature. . .

More directly and to the point they could only have been deposited after death entered the world, which means after man had *sinned*!

They were. . .formed under. . .the same basic physical *laws* that now exist but. . .not. . .at the same *rates* as at present.

. . .these processes (sedimentation, erosion, volcanism, tectonism, radioactivity, glaciation etc.) must have operated at greatly augmented rates and over greatly enlarged areas.'

This is the sort of nonsense Creationists demand the right to teach in public schools under the guise of 'Flood Geology'.

We have a remarkably complete fossil record, from Cambrian times to the present, especially of organisms with hard parts. The sequence in which major groups of animals and plants appear in and disappear from the geological record is now well known and meticulously documented in scientific literature. Since the time of William Smith, fossils have been, and still are, the best and most accurate method of establishing the sequence of deposition of rocks in which they occur. The orderly sequence of fossils in the rocks is a *fact*, readily accessible to anyone wishing to check it. All you need is a hammer, a good pair of legs and a keen, enquiring mind.

If there were *any* truth in the Creationist case, the scientific interpretation of the fossil record would be the easiest thing in the world to disprove. Any competent palaeontologist can suggest dozens of fossil forms which would, if discovered out of sequence in geological time, cause a reassessment of our understanding of the development of life on Earth. The fact that Creationists have been unable to come up with a convincing refutation, in over 200 years of trying, merely underlines the poverty of their arguments.

THE AGE OF THE EARTH — SCIENTIFIC VERSION

One of the great achievements of modern science has been the demonstration, based on information from several independent sources, that the Earth is extremely old — around 4.5 to 4.6 thousand million years — a far cry from the 6–10 thousand years claimed by Creationists.

The fossil record does *not* directly tell us how old or how young a particular formation is in terms of millions of years. It provides us with a *relative* scale. It does it so well that it provides an invaluable working tool, used daily by thousands of palaeontologists employed by multinational corporations investing billions of dollars in the search for mineral resources.

The gradual realisation of the enormous age of the Earth did *not* come about initially because of any theory of evolution, it predated Darwin by several generations. Widespread excavations exposed spectacular cross-sections of the rocks underlying the landscape and brought about the realisation that a very long and complex sedimentary history was involved.

Since the eighteenth century, many attempts have been made to calculate

TABLE 5.2
Estimates of age of the Earth

Date	Author	Age of earth	Basis of calculation
ca.1600	W. Shakespeare	'The poor world is almost six thousand years old' (*As You Like It*, Act IV Scene I)	
1642	J. Lightfoot	3928 BC	Biblical chronology
1650	Bishop Ussher	4004 BC	Biblical chronology
1778	James Hutton	'no vestige of a beginning — no prospect of an end'	Empirical observations
1778	de Buffon	75 000 years	Cooling rate of iron balls
1860	John Phillips	96 Ma*	Cumulative thickness of sedimentary strata
1863	Lord Kelvin	< 400 Ma	Cooling rate of solid body the size of Earth
1868	Lord Kelvin	100 Ma	Cooling rate of solid body the size of Earth
1868	James Croll	< 100 Ma	Glacial theory
1868	Charles Lyell	240 Ma since start of Cambrian	Revolutions of life forms since Cambrian
1876	Lord Kelvin	50 Ma	
1878	T.M. Reade	600 Ma	Rate of denudation of crustal sediments
1881	Lord Kelvin	20–50 Ma	
1897	Lord Kelvin	24 Ma	
1899	J. Joly	80–90 Ma	Ocean salinity and rates of supply from land
1895	Becquerel	discovery of radioactivity	
1907	B. Boltwood	> 2200 Ma	Radiometrically dated mineral samples
1931	A. Holmes	> 1600 Ma to < 3000 Ma	Radiometric dating of rocks
1946	Holmes and Houtermans	2900 Ma	Radiometric dating of rocks
1973	Patterson and Houtermans	4500 Ma (± 300 million)	Radiometric dating of rocks
1970s	Morris/Gish	6000–10 000 years	Biblical chronology

*Ma = 1 million years

a scale of 'absolute ages', expressed in years or millions of years, for various features of the geological record. To Hutton and later uniformitarians, the minimum age had, clearly, to be expressed in many millions of years. Various nineteenth-century estimates were based on the thickness of sediments, erosion rates of continents, cooling rates of a solid body the size of Earth, salinity of the oceans and so on. There were too many unknown and/or variable factors involved for accurate estimates to be made. The most influential nineteenth-century estimates were undoubtedly those of Lord Kelvin, one of the dominant figures in physics in the latter half of the century. He assumed that the Earth had cooled from an original molten state and, basing his estimate on known rates of cooling of large solid bodies, came up with estimates for the age of the Earth ranging from 400 million years (in the 1860s) to as little as 24 million years (in the 1890s). His calculations assumed that no new, or unknown source of, heat was present inside the planet.

The situation changed dramatically in the 1890s with Becquerel's discovery of radioactivity. Since then a vast amount of information has become available on the nature and natural occurrence of radioactive elements, partly because of our need to detect and measure radioactivity in an age of nuclear reactors and weapons.

In the early twentieth century it was realised that the long life spans and remarkably steady decay rates of radioactive elements provided an invaluable means for directly dating rocks. At the same time, it was realised that natural radioactivity provided a previously unrecognised source of internal heat in the Earth, completely changing ideas of the processes going on in the planet's interior.

By the 1920s, a clear picture began to emerge of the magnitude of the age of the Earth — thousands of millions of years. A list of attempts to calculate the Earth's age is given in Table 5.2. Note how the estimates of age have increased since Bishop Ussher's calculation in 1650. Note also that today's Creationists use essentially the seventeenth-century calculations of Lightfoot and Ussher, at the same time trying to justify their Biblical arguments by pseudoscientific re-evaluation of various scientific estimates which show an ancient planet.

RADIOMETRIC DATING

Radioactive elements have unstable nuclei which progressively decay to more stable daughter elements, emitting radiation and particles in the process. Each radioactive element has a distinctive, virtually invariable, decay rate. Different elements have different decay rates, expressed in terms of half-lives (the length of time it takes for half of the atoms in a certain quantity of the parent element to decay to the daughter element). Because the nuclear forces binding sub-atomic particles together are extremely powerful over the short distances involved, decay rates are virtually

TABLE 5.3
Elements used in radioactive dating

Radioactive isotope	Half-life	Daughter isotope	Use in geology
Uranium 238	4510 Ma*	Lead 206	only over 20 Ma
Uranium 235	713 Ma	Lead 207	only over 20 Ma
Thorium 232	13 900 Ma	Lead 208	only over 50 Ma
Potassium 40	1300 Ma	Argon 40	only over 100 000 years
Rubidium 87	47 000 Ma	Strontium 87	only over 10 Ma
Carbon 14	5570 years	Nitrogen 14	not beyond 70 000 years

*Ma = 1 million years

unaffected by normal physicochemical processes.

Various parent–daughter isotope pairs are used in radiometric dating (see Table 5.3), each in different ways and applied to particular sets of geologic circumstances. It is often possible to use several, independent, methods on the same rock. When these methods yield similar results, within a small margin of experimental error, one can be sure the results are reliable.

Radiometric dating was not widely accepted as a reliable method until the early 1930s. Holmes (1937, 1960) published the first widely accepted time scales covering the period from the beginning of the Cambrian to the present. Harland *et al.* (1982, Frontispiece) have produced a compact, authoritative and surprisingly readable account — *A geologic time scale* — explaining in detail the techniques used and how they have developed over the past 200 years. The work summarises the philosophy, problems, terminology and calibrations involved in Earth dating. This book is strongly recommended to anyone interested in the subject.

Just how little the Phanerozoic (Cambrian to Recent) time scale has changed over the past fifty years can be judged from Table 5.4.

The modern geologic time scale is a synthesis of two quite different scales, each supported by a wealth of data. The so-called chronostratigraphic scale is a hierarchical division based on the sequence, seen in the field, of rocks and fossils. It is a non-linear, 'relative' scale, accepted internationally because of proven value in deciphering Earth history. The so-called chronometric scale is based on radiometric dating and on the predictable decay of naturally-occurring, long-lived, radioactive isotopes in the rocks. It is linear, 'absolute' and uses the standard international time unit, the second, although usually expressed in millions of years.

For a good introduction to the techniques of radiometric dating see Clark and Cook, *Perspectives of the Earth* (1983). The chapter on absolute time is available separately from the Australian Academy of Sciences for use in projects.

TABLE 5.4
Various versions of the geological time scale
showing time to beginning of periods in million of years

	Holmes 1937	Holmes 1960	Harland *et al.* 1982
Tertiary	68	68–72	65
Cretaceous	108	130–140	144
Jurassic	145	175–185	213
Triassic	193	220–230	248
Permian	227	265–275	286
Carboniferous	275	345–355	360
Devonian	313	390–410	408
Silurian	341	430–450	438
Ordovician	392	485–515	505
Cambrian	470	580–620	590

THE AGE OF THE EARTH — CREATIONIST VERSION

Although Biblical chronologies indicate that Noah's Flood took place some 1600 years later than the Creation, which would place it around 2400 BC, it should be noted that various modern Creationist texts hedge on this point. Whitcomb and Morris (1961b: 489) concluded that: 'While Genesis II need not be interpreted as a strict chronology', at the same time 'it would seem to us that even the allowance of 5000 years between the Flood and Abraham stretches Genesis II almost to the breaking point.'

It is not uncommon to find some modern fundamentalists prepared to accept a Creation date around 10 000 years ago, about 8000 BC, but this is usually as far as they are prepared to go.

Morris (1972: 94): 'The only way we can determine the true age of the earth is for God to tell us what it is. And since He *has* told us, very plainly, in the Holy Scriptures that it is several thousand years in age, and no more, that ought to settle all basic questions of terrestrial chronology.'

God may have told Henry Morris, very plainly, his justification for such an all-encompassing assertion. For those of us reliant on the King James version of the Bible, the only specific age mentioned is Bishop Ussher's seventeenth-century marginal addition. Otherwise, the Bible is remarkably non-specific on the subject.

Morris (1974b: 136): 'For those who believe in Creation, therefore, physical processes and evidence that indicate an immense time scale must be explained away. Only those processes or evidence commensurate with a short (i.e. 6000 years) time scale can be accepted for use in creationism.'

On the Creationists' own admission, to compress the geological record into 6000 years requires a belief in catastrophes. The converse is *not* true!

Morris (1974b: 136):

'As a matter of fact, the creation model does not, in its basic form, *require* a short time scale. It merely assumes a period of special creation sometime in the past, without necessarily stating when that was. On the other hand, the evolution model does *require* a *long* time scale. The creation model is thus free to consider the evidence on its merits, whereas the evolution model is forced to reject all evidence that favours a short time scale.

Although the creation model is not necessarily linked to a short time scale, as the evolution model is to a long time scale, it is true that it does fit more naturally into a short chronology. Assuming the Creator had a purpose centred primarily in man, it does seem more appropriate that he would not waste aeons of time in essentially meaningless caretaking of an incomplete stage or stages of His intended creative work.'

Morris (1963: 56) goes even further in spelling out the implications of his interpretation of creation:

'No true creation is now taking place in the world and this revelation is confirmed by the great principle of mass and energy conservation.

Now this can only mean that, since nothing in the world has been created since the end of the creation period, everything must *then* have been created by means of processes which are no longer in operation and which we therefore cannot study by any of the means or methods of science. We are limited exclusively to divine revelation as to the date of creation, the duration of creation, the method of creation and every other question concerning the creation. And a very important fact to recognize is that true creation *necessarily* involves creation of an 'appearance of age'. It is impossible to imagine a genuine creation of anything without that entity having an appearance of age at the instant of its creation. It would always be possible to imagine some sort of evolutionary theory for such an entity, no matter how simple it might be, even though it had just been created.'

Morris's 'appearance of age' argument to explain away the antiquity of the Earth derives directly from a notorious nineteenth-century attempt at solving the problem. In '*Omphalos*' (Greek for navel), Philip Gosse (1857) suggested that Adam must have had a navel, *as if* he had been born of woman! Similarly, trees were created with rings in their wood *as if* they had been growing for years. Fossils were created buried deep in rocks *as if* they had been actual animals and plants which had once lived and died....

The implication that fossils had been put in the rocks by a creator, merely to test the faith of man, severely tested the credulity and tolerance of totally committed nineteenth-century Christians. As the 'appearance of age' concept involves not only the instantaneous creation of the Earth but also of the whole *universe*, with an apparent, but false, age the degree of deceit required on the part of such a creator ensured that Gosse's ideas were received with ridicule, not only by scientists but by leading clergymen such as Charles Kingsley, a vigorous opponent of evolution.

To find Morris resurrecting the '*omphalos*' argument in the Creationist cause in the late twentieth century is bizarre in the extreme and suggests a measure of desperation in trying to prove his case.

RADIOMETRIC DATING — CREATIONIST VERSION

The fact that early, rough estimates of the age of rocks, based on a variety of methods, have been subsequently confirmed, and the time scale vastly increased, by modern radiometric techniques does not disturb Creationists. They simply question the whole process without offering serious, testable, alternative hypotheses. They counter the scientific results by claiming that there is no certainty that radioactive decay rates have always been constant through geological time, and maintain that such rates *must* have been greater in the past (see discussion in Brush (1982)). They apply the same approach to *all* physical constants that pose problems for a young-Earth model (see Setterfield, 1981; refuted by Fackerell, 1984).

Gish (1973: 42): 'The interpretation of geologic data according to flood geology would include an evaluation of all dating methods, including especially a critical review of radiometric dating methods.'

Morris's (1974b: 133) superb example of double-think would be hard to beat:

> '*Rocks are not dated radiometrically*. Many people believe the age of rocks is determined by study of their radioactive minerals — uranium, thorium, potassium, rubidium etc. — but this is not so. The obvious proof that this is not the way it is done is the fact that the geological column and approximate ages of all the fossil bearing strata were all worked out long before anyone ever heard or thought about radioactive dating...[1974b: 137] Not even uranium dating is capable of experimental verification, since no-one could actually watch uranium decaying for millions of years to see what happens.'

What a curious inverted form of logic to justify a belief!

CONTINENTAL DRIFT AND PALAEOMAGNETISM

The idea that continents have moved relative to each other through time, and evidence for such movement, was expounded by Alfred Wegener between 1915 and 1930. In the absence of any then-known convincing physical mechanism for moving continents through (or over) the ocean floor, Wegener's theory was not widely accepted, although it was kept alive by some eminent geologists. Evidence that continents have moved, and are moving, is now overwhelming — we can even measure the rate of movement in the short term — and comes from a number of different sources.

The rocks of the Earth's crust retain many different kinds of information about their origin and history. One of the most useful is palaeomagnetism — a fossil record of the planet's magnetic field at the time of formation of the rock. Because the magnetic and geographic poles generally lie fairly close together, determination of the position of the magnetic poles of the past also provides a guide to the probable position of the contemporaneous geographic poles.

Studies of the palaeomagnetism of samples of continental volcanic rocks of different ages have demonstrated that the magnetic poles have changed position over time, appearing to 'wander' widely over the face of the Earth. It is now realised that in most, but not all, cases it was the continents, not the magnetic poles, which moved. The rocks of each continent preserve a magnetic record over time of the position of that continent relative to the poles.

For example, the 'polar wander paths' for Africa and South America, plotted independently, follow similar but widely separated tracks (Figure 5.1a). However, if the continents are brought together as shown in Figure 5.1b, the two separate wander paths merge. The obvious conclusion is that for the period represented by the magnetic records (dated radiometrically and/or palaeontologically as covering the period 450–200 million years ago) Africa and South America were closely joined, moved as one unit and began to separate 200 million years ago.

Palaeomagnetic studies reveal another phenomenon — reversals in polarity of the magnetic field. The polarity of Earth's magnetic field may change rather suddenly, for reasons still poorly understood. At intervals of hundreds of thousands to several million years, again dated radiometrically, the north and south magnetic poles 'flip' or reverse polarity, remaining thus for a considerable time before returning to the original polarity. A record of hundreds of such polarity reversals is preserved in iron and titanium minerals of igneous rocks.

In the early 1960s it was discovered that the basalts of the ocean floor display a striking pattern of magnetic anomalies, detectable by surface equipment, lying in stripes parallel to the seismically and volcanically active mid-ocean ridges. The magnetic stripes on either side of a ridge show a mirror-image pattern (Figure 5.2). It was soon realised that new oceanic crust, formed by magma, rising along the crest of a ridge in the axial rift valley, took up the prevailing north–south or south–north magnetic imprint as it cooled.

This new crustal material splits earlier crustal rock along the line of the ridge and moves it apart. The new material is, in turn, split apart by upwellings of new magma. The magnetic reversal patterns in the ocean-

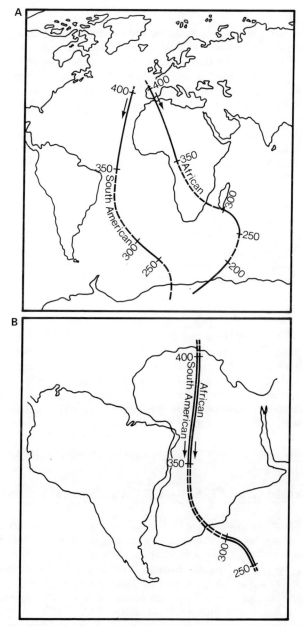

Figure 5.1. Palaeomagnetic record of the change with time of the position of the Earth's magnetic poles for Africa and South America — the 'polar wander paths': (a) with the continents in their present positions and (b) with the continents reassembled into their Gondwanaland position.

floor basalts, coupled with radiometric and biostratigraphic dating of the basalts, provide an accurate method of estimating the rate of sea-floor formation at the ridge, or the spreading rate on either side of the ridge. Present oceanic spreading rates vary from 2–3 cm/year in the North Atlantic and 3–4 cm/year in the South Atlantic to 10–18 cm/year along the East Pacific Rise. Sea-floor spreading between Australia and Antarctica is about 7 cm/year (Figure 5.3).

Additional evidence for the reversals comes from the deep oceans. These receive very little sediment from the continents, but sediments made up of the fossils of minute marine creatures accumulate slowly. Cores of ocean sediments, recovered during the Deep-Sea Drilling Program at several hundred sites in the world's oceans, retain a palaeomagnetic record of polarity reversals. Rates of sedimentation may have varied from site to site, but the same sequences of polarity reversals can be recognised in cores taken half a world apart. The deeper cores passed through the sediments to the underlying basalts, yielding samples of them for radiometric dating and correlation with magnetic anomalies detected by surface equipment.

Spectacular confirmation of the movement of continents about the Earth's crust comes from the use of space-age technology. Direct measurements of continental drift are currently coming from two quite independent methods of investigation (see *New Scientist*, 31 May, 1984: 6). Radio telescopes using very long baseline interferometry and laser signals bounced off satellites equipped with suitable reflectors can measure the separation of widely separated places on Earth with an accuracy of a few centimetres. Data already available confirm estimates of the rate and direction of crustal movement derived from geologic studies. The measured and calculated rates are in cm/year, yet thousands of kilometres of oceanic crust have been generated and continents have moved similar distances. The Earth must, therefore, be old. There is no way to accommodate such data in a young-Earth model as required by Creationists.

PALAEOMAGNETISM — CREATIONIST VERSION

Few of the 'scientific' Creationist accounts even mention the palaeomagnetic reversal story because it is clearly incompatible with a young-Earth interpretation. Instead, Creationist accounts (Morris, 1974b) repeat an idea originally proposed by Thomas Barnes (1973), a leading Creationist and Professor of Physics at the University of Texas, El Paso. Noting that the magnetic field strength of the Earth has decreased noticeably over the past 135 years, a fact not disputed by physicists, Barnes claimed that the magnetic field is decaying exponentially with a 'half-life' of 1400 years! This would mean, according to him, that the magnetic field was twice as strong 1400 years ago and thirty two times as strong 7000 years ago.

As Morris (1974b: 157) remarked, it is '. . . almost inconceivable that it

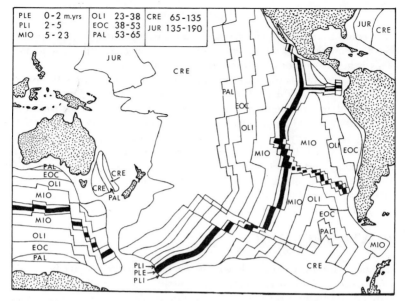

Figure 5.2. Pattern of the geological record on each side of the mid-oceanic ridges of the Pacific and Southern Oceans as disclosed by the palaeomagnetic record. PLE, Pleistocene; PLI, Pliocene; MIO, Miocene; OLI, Oligocene; EOC, Eocene; PAL, Paleocene; CRE, Cretaceous; JUR, Jurassic.

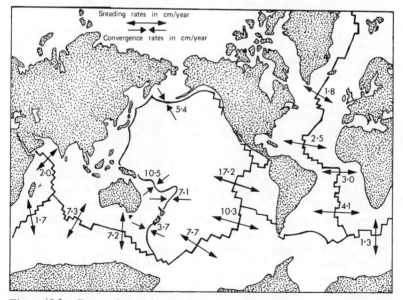

Figure 5.3. Rates of spreading and convergence of the oceanic floor. The direction is shown by arrows.

ever could have been much stronger than this. Thus 10 000 years ago the earth would have had a magnetic field as strong as that of a magnetic star! This is highly improbable, to say the least.'

Barnes's extrapolation from the present completely ignores the overwhelming geophysical evidence, from every continent and ocean floor, that the Earth's magnetic field not only fluctuates in intensity but that it has reversed its polarity on numerous occasions over the past 200 million years. Dalrymple (1982, 1983), Brush (1982) and other physicists have shown that Barnes's decaying magnetic field model is based on false assumptions and is contrary to what is known about the history and behaviour of the magnetic field through geological time. In spite of this, Barnes's travesty of a theory is still widely cited by Creationists as one of the best and most conclusive proofs of a very young Earth.

PLATE TECTONICS

The ocean floor is geologically young. The oldest rocks we have been able to recover from the ocean floors are Jurassic — a mere 180 million years old. The Earth itself is some twenty-five times older than the oldest ocean floor, and most of the ocean floor is much younger even than this. Some process is in operation which completely recycles the entire ocean-floor crust in 180 million years or so, about 4% of the total age of the planet.

The theory of plate tectonics accounts for this and many other observations.

Since the 1960s it has become increasingly clear that the Earth's crust is divided into a number of large and small 'plates' of oceanic crustal material which are in constant motion relative to each other. They separate along mid-ocean ridges where new ocean floor is generated. They collide along deep-sea trenches, volcanic island arcs and some continental collision zones. They slide past each other such as along the San Andreas Fault in California. The lighter continental masses ride passively on the oceanic crustal plates and are carried apart or driven together in various combinations.

Evidence for sea-floor spreading and the means of dating it and measuring its rate come from radiometric dating of new and older ocean crust, mirror-image lateral separation of magnetic reversal anomalies along ridges, vertical separation of magnetic reversal anomalies in deep-ocean sediments and the microfossil sequences in the same sediments. We have a coherent picture of new oceanic crust being formed continuously by slow, steady processes acting over millions of years.

The world-girdling system of mid-ocean ridges with their axial rift valleys is now extremely well known. Techniques of seismic detection, stimulated and refined by the need to monitor a nuclear test-ban treaty, have revealed a fascinating pattern of earthquake activity. Earthquakes are overwhelmingly confined to narrow zones which coincide with the mid-ocean ridges but also follow the lines of numerous deep-sea trenches and their associated volcanic island arcs and mountain belts with active volcanoes. Clearly, some mechanism keeps the Earth in a dynamic, changing state.

Over the past 20–25 years, with the development of the plate tectonics theory, physical geology texts have been rewritten so thoroughly that many pre-1960 texts are of little more than historical interest. The theory of plate tectonics has had remarkable success in incorporating and explaining a vast number of data from different sources and has stimulated many new lines of research. There are many comprehensive popular accounts of the theory (Miller, 1983; Smith, 1982a; Stevens, 1980; Sullivan, 1974). Readers requiring more information should consult one of these.

PLATE TECTONICS — THE CREATIONIST REACTION

During the last twenty-five years that the Earth sciences have been in almost constant revision, startling new discoveries and theories to explain them have been announced almost monthly. These have been closely scrutinised and evaluated. Many have been rejected, others accepted after modification. The plate tectonics theory has been developed and extended to accommodate a wealth of new information.

It has been an exciting and stimulating period to be working in the Earth sciences.

During the same period, the Creationists have been most active and vocal. A look at Creationist literature over the same period, however, shows virtually no mention of the flood of new data which led to the gradual acceptance of the idea of continental drift, sea-floor spreading and the development of the plate tectonics theory which unites all the findings.

The contemporary Creationist version of Earth history, trapped within the strait-jacket of a young-Earth, Flood Geology model, little changed from the eighteenth century, obviously cannot cope with, or face up to, the influx of new data from many different sources. Creationists cannot admit that the Earth could be millions of, far less thousands of millions of, years old. Because they have painted themselves into a corner they have no right to insist that we join them there!

Whitcomb and Morris (1961b: XXVII) in their preface to *The Genesis Flood* were remarkably honest (or naive): 'As we have stressed repeatedly in our book, the real issue is not the correctness of the interpretation of various details of the geological data, but simply what God has revealed in His Word concerning these matters. This is why the first few chapters and the two appendixes are devoted to a detailed exposition and analysis of the Biblical teachings on creation, the Flood and related topics.'

They went on to complain that geologists do not deal with actual Biblical evidence: 'The only conclusion that one can draw from this is that the authors and their critics seem to be operating on two entirely different sets of suppositions. On the one hand, scientific data are interpreted in the light of Biblical revelation; on the other hand, both revelation and the scientific data are interpreted in the light of the philosophic assumptions of uniformity.'

I could not have expressed it better myself. I was reminded of statements by two distinguished religious philosophers:

'I must believe in order that I may understand.'
St Anselm (1033–1109)
'I must understand in order that I may believe.
By doubting we come to questioning
and by questioning we perceive the truth.'
Peter Abelard (1079–1142)

EPILOGUE

In this account little or no mention has been made of Darwin, Darwinism, natural selection, or evolution. The omissions are intentional. The geological record stands on its own. Much of it is investigated by scientific disciplines far removed from, and quite independent of, those used in the biological sciences. The theory of evolution may be strongly supported by the fossil record of life preserved in rock; interpretation of the record of the rocks is not dependent on a belief in the theory of evolution. Fundamentalist Creationists endeavour to disguise that fact when they attack the geological sciences.

The choice is simple. It is between a scientific interpretation of Earth history based on empirical observation and a fundamentalist religious interpretation based on faith. 'Now faith is the substance of things hoped for, the evidence of things not seen' (11th Chapter of St Paul's Epistle to the Hebrews, New Testament, King James Version) but it is not science.

FURTHER READING

Scientific

For the most comprehensive list of source material on Creationism as applied to geology, see Shea, J.H. (1984) A list of selected references on Creationism. *Journal of Geological Education* 32: 43–49.

Chorlton, W. (1983) *Planet Earth — ice ages.* Amsterdam. Time-Life Books. 176 pp.

Craig, G.Y., McIntyre, D.B. & Waterston, C.D. (1978) *James Hutton's Theory of the Earth: the lost drawings.* Scottish Academic Press, Edinburgh. 68 pp.

Futuyma, D.J. (1983) *Science on trial — the case for evolution.* Pantheon Books, New York. 251 pp.

Geological Museum, London

The museum has produced a range of inexpensive, non-specialist, well-illustrated booklets of 36 pages. These are ideal for schools. They include:

The story of the Earth (1972);
Volcanoes (1974);
Moon, Mars and meteorites (1977);
The age of the Earth (1980).

Gillispie, C.C. (1951) *Genesis and geology — the impact of scientific discoveries upon religious beliefs in the decades before Darwin*. Harper, New York. 306pp.

Holmes, A. (1965) *Principles of physical geology*. (2nd edition) Nelson, London. 1288 pp.

Hsu, K.J. (1972) When the Mediterranean dried up. *Scientific American* 227(6): 26–36.

Kitcher, P. (1982) *Abusing science — the case against Creationism*. Massachusetts Institute of Technology Press, Cambridge, Mass. 213 pp.

Montagu, A. (ed.) (1984) *Science and Creationism*. Oxford University Press, Oxford. 430 pp.

Moyer, W. (1983) Evolution and young-Earth creationism. *In* The Creationist attack on science — symposium. 1982. *Proceedings of the Federation of American Societies for Experimental Biology* 42(13):3025–3030.

Owen, H.G. (1981) Constant dimensions or an expanding Earth? *In* Cocks, L.R.M. (ed.) *The evolving Earth*. British Museum (Natural History), London: 179–182.

Ruse, M. (1982) *Darwinism defended — a guide to the evolution controversies*. Addison-Wesley, Reading, Mass. 356 pp.

Shea, J.H. (ed.) (1983) *Journal of Geological Education 31*(1): 1–58 (Seven articles on geology and Creationism.)

Shea, J.H. (ed.) (1983) *Journal of Geological Education 31*(2): 72–144. (Eleven articles on geology and Creationism.)

Creationist

Coffin, H.G. (1969) *Creation — accident or design?* Review and Herald Publishing Association, Washington. 512 pp.

Morris, H.M. (1977) *The scientific case for Creation*. Creation-Life Publishers, San Diego, Calif. 87 pp.

Slusher, H.S. & Gamwell, T.P. (1978) *The age of the Earth*. Institute for Creation Research, San Diego, Calif. 77 pp.

CHAPTER SIX

EVIDENCE FOR EVOLUTION FROM THE FOSSIL RECORD

Michael Archer

We *have* found links between modern humans and fossil apes...In
many cases, the problem is not
a lack of intermediates but the
existence of so many closely related
intermediate forms that it is notoriously difficult to decipher true
ancestral-descendant relationships.
In a very real sense, the fossil record
is far better testimony to evolutionary change than Darwin, in his later
years, probably imagined possible.
(Godfrey, 1983: 199)

INTRODUCTION

We have already considered some of the evidence for the reality of
evolution that has come from study of living organisms (Chapter 3). This
unequivocally demonstrates the existence of processes that are leading to
evolutionary change at the present time. To determine if these processes
operated in the past requires evidence sought in the Earth's own filing
system: the fossil record. This record can never be perfect because not all
organisms lived in situations conducive to fossilisation and because most of
the record that did accumulate has either been removed by erosion or is
too deeply buried for discovery. There should, however, be enough of the
record remaining and accessible to test whether or not any patterns it
displays support or refute the predictions of the evolutionary hypothesis.

While we have no difficulty demonstrating the existence of small-scale
evolutionary changes, such as species-to-species transitions, in modern
organisms, the fossil record should enable us to test whether the same sort
of changes occurred in the past. It should also allow us to test hypotheses
about higher-level evolutionary relationships such as family-to-family or
class-to-class transitions. This second type of hypothesis is difficult to test
using living organisms because the time required for changes of this

magnitude would prohibit direct observation. As we have seen, the only way to do it using living organisms is indirectly, for example by examining the evidence of comparative molecular biology. What does the fossil record have to offer in the way of tests of these evolutionary predictions?

THE FOSSIL RECORD OF SPECIES-TO-SPECIES TRANSITIONS

It is probable that much of the lack of evidence for transitional forms between fossil species stems from the incompleteness of the fossil record. Palaeontologists are, however, steadily filling in the missing bits as new discoveries are made and today a vast amount more is known than in Darwin's day. Moreover, the rate of accumulation of knowledge about fossils is now greater than it ever has been. For example, the finding in 1983 of just one new small area of Tertiary mammal-containing deposits in Australia has more than doubled our previous total knowledge about the Tertiary mammals of the continent. It would be premature to say what tomorrow's discoveries may reveal about evolutionary history.

We do not have to wait until tomorrow for demonstrations from the fossil record of the reality of species-to-species transitions. There are hundreds of studies demonstrating phylogenetic sequences at the species level in the fossil record (see Further Reading).

Gradual Transitions Between Extinct Species

Collection of fossil primates — the group to which humans, monkeys, prosimians and a large variety of other forms belong — from Early Tertiary deposits in Wyoming has been going on since the 1800s. The early collections consist mostly of single specimens from single localities. However, over the years, palaeontologists have accumulated very large collections from these sediments, including many specimens of prosimian primates belonging to the genera *Cantius, Pelycodus* and *Notharctus*.

Recently, Gingerich and his palaeontological colleagues (see Further Reading) collected intensively from hundreds of sites in these Eocene sediments which are more than 500 m thick, representing approximately five million years of uninterrupted accumulation. They then arranged the fossil specimens from the sediments in stratigraphic sequence, studied the specimens from each successive level and sorted them into morphological categories. They also studied specimens from earlier collections.

When samples of particular species, such as those belonging to the genus *Cantius*, were measured and the results plotted against stratigraphic position a distinct pattern emerged (Figure 6.1) that strongly suggests progressive phyletic evolution through time. For example, *C. mckennai* clearly merges, through small changes in size and shape, into *C. trigonodus*, the next species represented higher in the stratigraphic sequence. Similar transitions occur further up the succession. Towards the top of the sequence there is

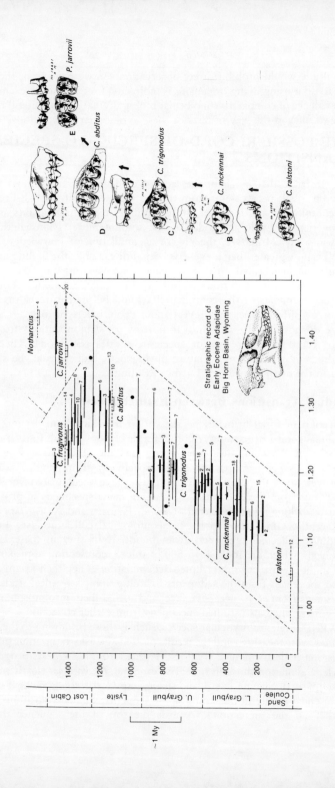

P. jarrovii

E

C. abditus

D

C. trigonodus

C

C. mckennai

B

C. ralstoni

A

Notharctus

C. jarrovii

C. frugivorus

C. abditus

C. trigonodus

C. mckennai

C. ralstoni

Stratigraphic record of
Early Eocene Adapidae
Big Horn Basin, Wyoming

1.00   1.10   1.20   1.30   1.40

1400

1200

1000

800

600

400

200

0

Lost Cabin   Lysite   U. Graybull   L. Graybull   Sand Coulee

~1 My

indication of a separation where ancestral populations descended from *C. abditus* underwent sufficient change to be recognised as species of a different genus, *Pelycodus*. This in turn appears to have been the ancestral stock that evolved into the species of the yet younger genus *Notharctus*.

The importance of this stratigraphic sequence is that, without a detailed knowledge of the fossil record, the named forms shown at the right of Figure 6.1 would appear to have been distinct species, lacking intermediate forms to demonstrate their origins. With a detailed knowledge of the fossil record, however, it becomes clear that populations intermediate in morphology occupy intermediate stratigraphic positions between the named forms. It is precisely the sort of example that Darwin predicted the fossil record would reveal, one in which species boundaries in time could only be defined by use of arbitrary distinctions.

This sort of pattern in the fossil record has also been demonstrated for other groups including some lineages of foraminiferans, ammonites, brachiopods, trilobites, molluscs, fish, reptiles and mammals (e.g. Martin, 1984). Gingerich (1976, 1984) has analysed several other sequences of Early Tertiary mammals demonstrating that interspecies transitions are evident when the fossil record is examined in enough detail. Wolpoff (1984) and Brace (1983) have shown that within the fossil record for species of our own genus *Homo*, detailed analysis reveals, for example, a gradual transition between the extinct *H. erectus* and the surviving *H. sapiens*. Transitional sequences (although less complete) also link *H. erectus* to *H. habilis* and the latter back to species of *Australopithecus*.

Non-gradual Transitions Between Extinct Species

While there is clear evidence for transitions between species through time, there is also a very spritely controversy amongst evolutionists about the tempo of these speciation events. The debate is not about whether the fossil record provides support for the hypothesis that younger species descended from older ancestral species but rather whether or not they did so by the steady accumulation of minor changes. For example, some palaeontologists have challenged Gingerich's interpretation of the *Cantius* sequences, claiming that despite the evidence for gradual evolutionary transition, the sequences are not able to demonstrate precisely how rapidly the speciation events occurred (Stanley, 1979).

In contrast to the phyletic 'gradualism' implied by Gingerich's studies,

◄ **Figure 6.1.** Results of a morphological study of *Cantius* spp. (Gingerich & Simons 1977; modified from Stanley, 1979). Thin horizontal lines represent size range for each set of specimens. Thicker lines represent standard error around the mean. Means indicated by short vertical lines. Black dots represent single specimens only. Horizontal dashes indicate samples from area adjacent to main study area. Representative specimens of upper molars and premolars shown at right. See text for full explanation.

where speciation is viewed as an inexorable consequence of steady divergence, there is an alternative hypothesis known as 'punctuated equilibrium' (i.e. that species normally evolve in fits and starts). According to this model, the period of origination would be very short (of the order of thousands to tens of thousands of years) and would be followed by a much longer 'stasis' period (commonly of the order of millions of years) during which the species would undergo very little morphological change. It has also been suggested by some punctuationalists (Williamson, 1981a,b) that speciation events follow disruption of genetic mechanisms that restrict expression of morphological variation. The nature of these hypothetical restriction mechanisms has been questioned, but it is nevertheless one of the most stimulating ideas to come from study of the fossil record.

One of the most thorough studies providing evidence for the punctuated equilibrium model is that of Williamson (1981a) on fossil molluscs from the Turkana Basin in northern Kenya. He measured fifteen features in each of 3300 fossil shells representing thirteen species lineages. The fossils were obtained from an almost continuous 400-m-thick sequence of Plio–Pleistocene lake deposits spanning the period between 1.3 and 4.5 million years ago. His analysis demonstrated that, within these molluscan lines, speciation events were relatively sudden, occupying a time span of 5000–50 000 years. They were preceded by relatively long periods of stability of about three million years. Transitional populations leading to the evolution of new species were characterised by 'pronounced developmental instability, (i.e. they exhibited a high degree of variability).

Some geneticists (Jones, 1981) have pointed out that the rapid rate of speciation demonstrated by the Turkana molluscs is not surprising. In 5000 to 50 000 years, the molluscs would have produced a number of generations roughly equivalent to that produced by a *Drosophila* colony in 1000 years. Now, we know that in such a *Drosophila* colony, considerable change occurs through the generations produced in only five years. In the extended time span of thousands of years, many more generations are highly likely to result in speciation.

However, it is the long periods of stasis observed in all thirteen of the Turkana lineages that were not predictions inherent in the neo-Darwinian model of evolution (although they also do not conflict with it) or observed during laboratory experiments. Perhaps the fossil record can demonstrate things about the mode of speciation that study of living organisms cannot? Anyway, this is one of the more controversial suggestions being made by some of the punctuationalists.

More recently, an even more elaborate version of the punctuated equilibrium hypothesis has been developed (Bakker, 1983, 1985). This has been used to reinterpret the *Cantius* evolutionary sequence noted earlier and proposes that what has happened here is neither gradual evolution of one population through time nor simple punctuated equilibrium speciation events of one species into another. Perhaps, Bakker (1985) suggests, the new and stratigraphically higher species (each of which clearly represents a descendant of the older form that occurs stratigraphically below it) developed in rapid punctuated fashion after a founding population left its Wyoming homeland. At some time after this, following an opportunity to disperse back to their ancestral homeland, the new and slightly different daughter species subsequently replaced its ancestral species wherever the two came into contact. Bakker suggests that in this way evolutionary speciation events evidenced in the fossil record may actually represent evolution by 'revolution'.

Like other punctuationalists, Bakker suggests that species, once they initially and rapidly diverge from their ancestral parent, are relatively immune to gradual transformation into anything significantly different until another speciation event occurs. This post-speciation stability is the period of stasis recognised by other punctuationalists. In these punctuationalist models, for significant adaptational change to occur, speciation events must occur first.

At present, palaeontologists persuaded by the evidence for punctuated equilibria are having another hard look at the whole of the fossil record. The new interest they show is in the old awareness of the many gaps in knowledge about particular transitional forms. If the evidence of the Turkana speciation sequences is representative, no more than 0.001% to 0.02% of the fossil record would be expected to show evidence of speciation events. Considered in this light, the commonly encountered situation of lots of fossil species but few intermediate forms is precisely what the punctuated equilibrium model of evolution predicts.

Such a 'punctuated' record appears to be demonstrated, for example, by the last five million years of the fossil record of elephantine elephants (Figure 6.2). Only two species have survived; the African elephant (*Loxodonta africana*) and the Asiatic elephant (*Elephas maximus*). Some lineages, such as that leading from *E. ekorensis* to *E. iolensis* via *E. recki*, appear to have evolved gradually. Others, such as *E. celebensis*, although closest in morphology and time to *E. planifrons*, seem to have speciated rapidly, without leaving detectable traces in the fossil record of transitional forms.

Clearly, a number of different hypotheses compete to explain the mechanisms involved in producing the species-to-species transitions seen in the fossil record. But each of these is in agreement that the transitions indicate evolutionary events. What is being debated is not whether evolution has occurred, but *how* it occurred.

FUNDAMENTAL LINKS IN THE CHAIN OF LIFE

The second type of evidence sought from the fossil record to test the concepts of evolution is that for evolutionary transitions linking apparently distinct higher taxa.

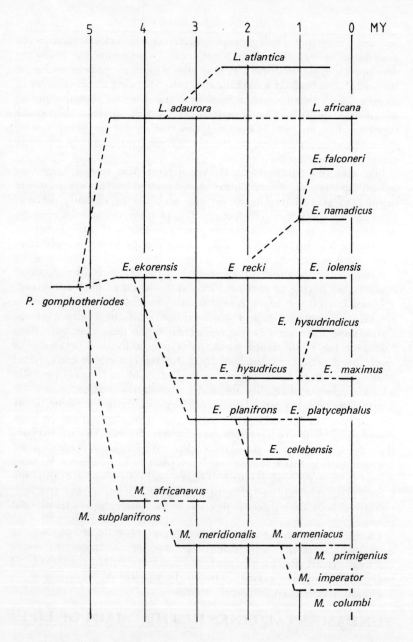

Figure 6.2. Fossil record of elephantine elephants. Solid horizontal lines represent fossil record; dashed lines represent phylogenetic interpretations about probable origins of the species. *E = Elephas*; *L = Loxodonta*; *M = Mammuthus*; *P = Primelephas*. (Archer & Clayton, 1984, after Stanley, 1979.)

Given the fact that erosional and depositional processes have been going on for at least 3 700 000 000 years, the age of the Earth's oldest known sedimentary rocks, one would predict not only gaps in the Earth's fossil record but also that their position would not be entirely random. They should occur more frequently and be of larger size in the older record because that is the part of the record that has had the longest time for removal by erosion, burial by younger sediments and/or destruction by metamorphosis. As a result, it comes as no surprise that continuous sequences of fossils are more difficult to find in rocks 500 000 000 years old than in rocks 50 000 000 years old. Nevertheless, there are thousands of known fossil forms that are intermediate in morphology and time between others. We will now consider a few extinct forms that bridge what otherwise appear to be significant gaps between major groups of modern organisms.

The Gap Between Living Reptiles and Birds

Amniotes are vertebrates that reproduce by means of an amniotic egg (see any modern zoology text). In conventional taxonomy, amniotes include reptiles (Class Reptilia), birds (Class Aves) and mammals (Class Mammalia). In the modern world, representatives of these three groups are quite distinct from one another and there is no problem in placing any living amniote in one of these three classes.

There is a fossil record of birds and bird-like amniotes (Feduccia, 1981) and the chapters on early birds in Archer and Clayton (1984). Of particular interest is *Archaeopteryx lithographica*. The first skeleton of this creature was found in 1861. The 140-million-year-old (Late Jurassic) lithographic limestone in which it was found also contained the skeletons of small carnivorous dinosaurs known as *Compsognathus longipes*. What made *Archaeopteryx* so strikingly different from these dinosaurs was the unmistakable impressions of feathers, the hallmark of birds. It displayed a few other features otherwise found only in birds, but most of its anatomy was reptilian and very similar to that of the small running dinosaurs with which it was found (see Table 6.1 and Figure 6.3). In fact, another skeleton of *Archaeopteryx* collected in 1855 was not recognised as such until 1970 because, lacking clear impressions of feathers, it had been assumed to be another skeleton of *Compsognathus*.

The five skeletons of *Archaeopteryx* now known have been the subject of many detailed anatomical studies. The conclusions are always the same; this is an animal that belongs in the structural and temporal gap between reptiles and birds. It would be difficult to imagine a more perfect link. If there has been any significant debate among palaeontologists about the evolutionary significance of *Archaeopteryx*, it is not about its intermediate position between birds and reptiles but rather about the particular group of reptiles to which it is most closely related (Feduccia, 1981; Thulborn, 1985a; Whetstone, 1983).

Figure 6.3. Skeleton of *Archaeopteryx lithographica* (a) compared with that of pigeon (b). See text and Table 6.1 for explanation. (From Molnar & Archer, 1984, after Colbert, 1969).

There is much more that is fascinating about the early evolutionary record of birds in terms of our present interest in 'links' (Feduccia, 1981; Molnar and Archer, 1984). For example, the Middle Cretaceous toothed birds (such as *Ichthyornis dispar*) represent a structural link between the Jurassic *Archaeopteryx* and the modern orders of birds, most of which made their first appearance in the Early Tertiary. These Cretaceous toothed birds display more avian features than did *Archaeopteryx* including a shortened tail, a synsacrum, a tarsometatarsus and uncinate processes on the ribs (see Table 6.1 for a brief definition of terms).

TABLE 6.1

Characteristics of living birds and living reptiles compared with *Archaeopteryx*. Definite characters of *Archaeopteryx* indicated by *. Other characters of *Archaeopteryx* indicated by *? (See foot of table for explanation)

Birds	Reptiles
Feathers *	Scales
Wings *?	No wings
Fused digits	Free digits (if limbs present)*
Fused phalanges	Free phalanges*
Hollow bones	Solid bones*
Sternum keeled	Sternum unkeeled*
Furcula[1] present*	Furcula absent
Synsacrum[2] present	Synsacrum absent
Backward-pointing pubis*?	Pubis not so
Very few tail vertebrae	Numerous tail vertebrae*
Pygostyle[3]	No pygostyle*
Uncinate (hooked) processes present on ribs	No uncinate processes*
Carpometacarpus[4] present	Carpometacarpus absent*
Tibiotarsus[5] present	Tibiotarsus absent*
Tarsometatarsus[6] present	Tarsometatarsus absent*
Skull bones almost entirely fused	Skull bones not so
Large brain and braincase *?	Small brain and braincase
Teeth absent in all forms	Teeth present in most forms*

[1] furcula (wishbone) = fused clavicles. Provides support for wings.
[2] synsacrum = fused pelvis and sacrum.
[3] pygostyle = last few tail vertebrae foreshortened and compressed.
[4] carpometacarpus = fused carpal and metacarpal bones.
[5] tibiotarsus = fused lower leg bone and ankle bones.
[6] tarsometatarsus = fused tarsal and metatarsal bones.
*? Feathered arms of *Archaeopteryx* probably functioned as wings. Brain was relatively large and braincase bird-like. Pubis was possibly backward-pointing. Tarsal bones were fused but not into a tibiotarsus.

Recently, a fascinating fossil bird that is intermediate in time between *Archaeopteryx* and the Middle and Late Cretaceous palaeognathous toothed birds has been described. *Ambiortus dementjevi*, an Early Cretaceous bird from Mongolia, is known from associated parts of a wing, shoulder girdle and vertebrae. Its pectoral apparatus shows features that are more like those of modern birds than the comparable conditions in the Late Jurassic *Archaeopteryx* while in other respects it closely resembles primitive palaeognathous birds recently discovered in the Paleocene and Eocene of North America and England (Emery and Schultze, 1984).

There is even a fossil record of feathers that demonstrates a structural gradient between simple reptilian scales and the complex feathers of *Archaeopteryx* (Molnar and Archer, 1984; Rautian, 1978). This fossil scale-to-feather transition is supported by studies of the embryology of the scales of living reptiles and the feathers of living birds. Feathers are a very specialised type of reptilian scale (Thulborn, 1985a).

The Gap Between Living Reptiles and Mammals

Being mammals ourselves, we tend to see the distinctions between the members of Class Mammalia and those of Class Reptilia as much more fundamental than those which separate fish from amphibia, amphibia from reptiles and birds from reptiles. Considering just modern mammals (because our interest here is in seeing whether the fossil record provides any forms that bridge significant 'gaps'), these differences seem to be very fundamental indeed.

Table 6.2 shows character pairs which separate modern mammals from modern reptiles. All of the structures listed, except the last six, involve parts of the skeleton and might, as a result, be expected to have some sort of fossil record. That being the case, if the evolutionary concept includes, as it does, the hypothesis that mammals evolved from reptiles (a hypothesis that was originally developed from study of living animals only), we might expect some evidence from the fossil record for transitional forms. Are they known? In grand abundance!

The fossil record shows evidence for a transition between reptiles and mammals. The evidence is based on detailed studies of jaw articulation in reptiles and mammals (both living and fossil) and the development of the bones of the mammalian middle ear, three of which appear to be the same as bones that help to articulate the reptilian jaw.

Soon after the 'true' (rather than transitional) reptiles had made their appearance, two distinctive subgroups of reptiles began to dominate the ecosystems of the day: the more conventional reptilian forms known as diapsids (such as dinosaurs, lizards and snakes) and the somewhat less conventional synapsids or mammal-like reptiles.

The oldest synapsids known are from the Carboniferous (i.e. about 315 000 000 years in age). They were small (50 cm) reptiles distinguished from others of their day most conspicuously by the presence of a temporal space in the cheek region of the skull. The subsequent 100 million years of synapsid evolution is represented by many thousands of fossils that document a number of different evolutionary trends within the group. Synapsids declined at the end of the Triassic, about 215 million years ago, at about the same time that primitive mammals made their first appearance in the fossil record. One synapsid group, however, survived until the Middle Jurassic. With extinction of this group the synapsids as such disappeared, although there is incontrovertible evidence from the fossil record that the first mammals were their descendants.

We can trace the development of mammalian characteristics in the synapsids by examining the skulls and other skeletal parts found at stratigraphic levels corresponding to successively younger times.

Let's begin with a look at a synapsid which lived during the Middle

TABLE 6.2

Character pairs separating living mammals and living reptiles

Mammals	Reptiles
Skeletal characters	
One bone in lower jaw	>One bone in lower jaw
Jaw–skull articulation between dentary and squamosal bones	Jaw–skull articulation between articular and quadrate bones
Three small bones in middle ear	One bone in middle ear
Teeth differentiated into incisors, canines, premolars and molars (unless secondarily modified)	Undifferentiated teeth
Teeth occlude to facilitate cutting	Teeth do not occlude
One set of replacement teeth	Teeth replaced throughout life
Secondary palate separating air and food passages	No secondary palate (except in crocodiles, where developed independently)
No prefrontal or postorbital bones in skull	Prefrontal and postorbital bones present
No postorbital bar	Postorbital bar present
Large openings present in sidewall of skull	If openings present, small
Two occipital condyles	One occipital condyle
Slender zygomatic arch arising low on the skull	Not in living reptiles
Fenestra ovalis confined to periotic bone	Fenestra ovalis differently constructed
Large ilium	Small ilium
Third and fourth digits of hand have three bones each	Third and fourth digits have four and five bones respectively
Relatively large brain	Relatively small brain
Non-skeletal characters	
Organ of corti developed from embryonic basillar papilla	Unelaborated basillar papilla in middle ear
Hair	Scales
Milk glands	No milk glands
Constant body temperature	Body temperature fluctuates with environmental conditions
Four-chambered heart	Three-chambered heart (except in crocodiles)
Diaphragm	No diaphragm (breathe by rib movements)

Triassic — *Probainognathus jenseni*, an animal clearly intermediate in structure and age between mammals and reptiles (Figure 6.4). This synapsid had *both* the mammalian dentary–squamosal jaw articulation and the reptilian quadrate–articular jaw articulation. No living reptile has the former and no

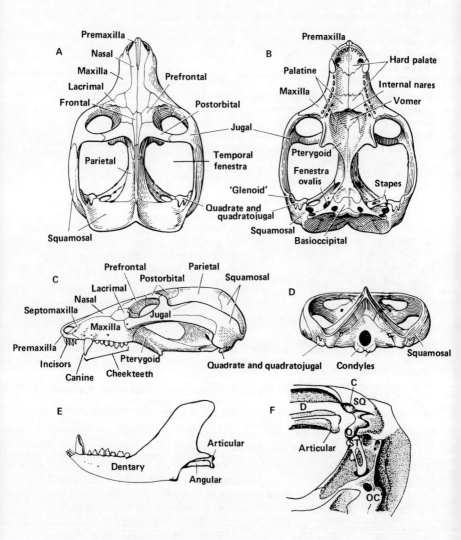

Figure 6.4. The skull (a–d), lower jaw (e) and jaw articulation (f) of the Middle Triassic *Probainognathus jenseni*, one of many extinct mammal-like reptiles intermediate in structure between living reptiles and living mammals. Like mammals it had a squamosal–dentary jaw articulation system but alongside this it also had, like reptiles, a quadrate–articular articulation system. *Abbreviations*: D, dentary; SQ, squamosal; C, area in squamosal where dentary articulates; Q, quadrate; ST, stapes (or 'stirrup'); OC, occipital condyles; ART, articular; ANG, angular; (a–d after Kermack and Kermack, 1984; e, f after Romer and Parsons, 1977).

living mammal has the latter form of jaw articulation (see Table 6.2). Although the quadrate and articular bones are much reduced they are still much larger than their present-day mammalian equivalents — the malleus and incus of the middle ear — into which, as we shall see, they have developed.

Other mammal-like features of *P. jenseni* include: the partially-developed secondary hard palate which is more developed than in reptiles but less so than in mammals; a double occipital condyle at the back of the skull; very large temporal fenestrae; teeth differentiated into incisors, canines and cheekteeth and just two generations of teeth. These features, in combination with others that are more reptilian (e.g. the presence of the prefrontal and postorbital bones — see Figure 6.4 and Table 6.2), place this and related synapsids squarely between other reptiles and mammals.

Younger synapsids such as *Diarthrognathus broomi* from the mid-to-Late Triassic showed additional mammal-like features. (Di — arthro — gnathus actually means 'two — jointed — jaw' if you know your Greek). These include loss of some of the 'reptilian' skull bones (e.g. the prefrontal and postorbital bones) which results in the joining up of the space for the eye socket and the temporal fenestra. The zygomatic arch was more slender (see Kemp, 1982, for detailed explanation).

Synapsids from the Late Triassic such as *Oligokyphus major* had strikingly mammalian skeletons. Their features are so mammal-like that there has been debate as to whether they should be regarded as reptiles or mammals.

All of these features found in synapsids are standard equipment in living mammals but unknown in living reptiles. In other features, such as the relatively small brain and poor differentiation of the cheekteeth row into molars as well as premolars, synapsids are more reptile-like than mammal-like. This mixture of mammalian and reptilian features in synapsids is precisely what one would expect to find in a group of organisms transitional between reptiles and mammals.

The history of synapsids demonstrates very plainly that they were a group of reptiles that, for the first time, was experimenting with mammalian features. From their beginnings in the Carboniferous, they display ever-increasing degrees of 'mammal-ness' until, by the end of the Triassic, some lineages were clearly on the borderline between more advanced synapsids and primitive mammals (Figure 6.5).

In fact, the questions being debated now amongst vertebrate palaeontologists are not whether or not mammal-like reptiles were structural and temporal intermediates between reptiles and mammals but rather which *particular* group of synapsids gave rise to mammals. And even here, debates (see Kemp, 1982) are not any wider than the choice between tritylodonts (e.g. *Oligokyphus*), tritheledonts (e.g. *Diarthrognathus*) or chiniquodonts (e.g. *Probainognathus*), all Middle-to-Late Triassic synapsids which were alive and kicking before the first primitive mammals appeared.

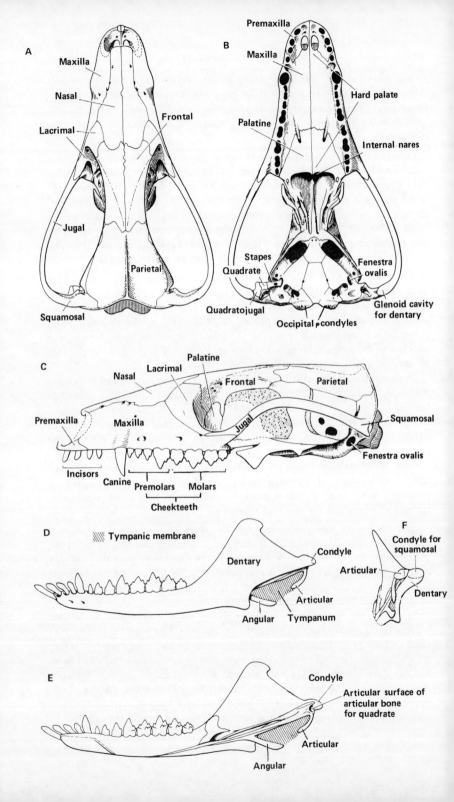

As noted above, the modern reptilian and mammalian jaw articulation systems are radically different. In the former, the articular bone of the lower jaw hinges with the quadrate bone of the skull. In modern mammals, the lower jaw consists of one bone, the dentary, which hinges against the squamosal bone of the skull. Skulls showing the synapsid-to-early mammal transition demonstrate that the mammalian condition was gradually acquired by posterior enlargement of the dentary bone until it contacted the squamosal bone (Figure 6.6).

The posterior extension of the dentary developed around the outside of the articular–quadrate system and (as the fossil record of synapsids and early mammals demonstrates), as the dentary–squamosal articulation enlarged, the articular–quadrate system diminished (Crompton, 1985).

There are several hypothetical explanations for why this transition occurred (Crompton, 1985; Kemp, 1982). Amongst those proposed are the following: a reduction in the number of bones of the lower jaw would reduce potential zones of weakness when the jaw was under stress; since the dentary held the teeth of the jaw it was the logical bone to enlarge; since other postdentary bones were successfully articulating the lower jaw, they continued to do so at least until the dentary–squamosal articulation was strong enough to cope with the task on its own; at first, the dentary–squamosal contact functioned more as a buttress than as a significant articulation system for the lower jaw; simultaneously the postdentary bones underwent selection for size reduction and eventually were either lost or (in the case of the angular and articular) became involved in other tasks taking place in the same region of the skull. Reasons for why the postdentary bones would have come under selective pressure to reduce in size involve the second aspect of this evolutionary transition: auditory reception.

From a detailed study of advanced synapsid skulls, Allin (1975) concluded that their tympanic membrane ('ear drum') occurred immediately in front of the articular–quadrate jaw articulation (which is where it occurs in some living reptiles such as amphisbaenids or worm lizards) suspended within the appropriately curved processes of the lower jaw bones (the angular and articular; see Figure 6.7). This meant that these bones and the small bones to which they were connected (the quadrate and stapes) which in turn made contact with the inner ear were involved in the transmission of sound *as well as* the task of articulating the jaw to the skull.

◀ **Figure 6.5.** The skull and lower jaw of one of the oldest known mammals, the Late Triassic *Eozostrodon parvus*. It displays a mingling of features that are found in mammals (e.g. the differentiation of the cheekteeth into premolars and molars, the widely extended parietal, the shape of the periotic which contains the fenestra ovalis, the slender jugal and the extent of posterior development of the hard palate) with those otherwise more typical of reptiles (e.g. the suite of postdentary bones still resting against the inside of the lower jaw). See the text for an explanation and Figures 6.7 and 6.8. (a–e after Kermack and Kermack, 1984.)

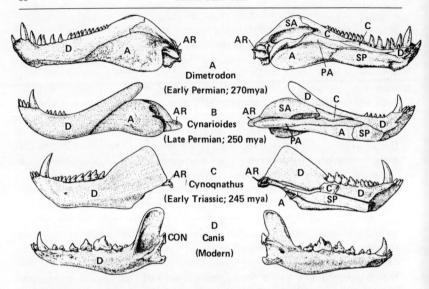

Figure 6.6. The left lower jaws of mammal-like reptiles (a–c) and a modern mammal, the dog (d), showing the progressive expansion of the dentary bone and the corresponding reduction in the postdentary bones. The illustrations on the left display the lateral (outside) views while those on the right display the mesial (facing the inside) views of the same jaws. This gradual trend started in Permian mammal-like reptiles (about 270 million years ago, mya) and culminated in Middle Jurassic mammals (about 175 mya) when all bones except the dentary were lost from the lower jaw although some were retained in the adjacent middle ear (see the text and Figures 6.7 and 6.8). *Abbreviations*: D, dentary; A, angular; AR, articular; SP, splenial; PA, pre-articular; SA, surangular; C, coronoid; CON, condyle on the dentary for articulation with the squamosal. (After Romer and Parsons, 1977).

Studies of living reptiles with these characteristics (amphisbaenids) have shown that the sound-transmission system interpreted by Allin as having been present in synapsids works quite well but that sensitivity to high-frequency sounds is definitely limited by the mass of the sound-conducting bones (Kermack and Kermack, 1984). Most *living* small mammals appear to depend on high sensitivity to high-frequency sound. This may have been no problem to larger synapsids, but to smaller ones might have been disadvantageous. It could, then, have resulted in the selective pressure that led to size reduction of the postdentary bones and their eventual freedom to function solely as sound transmitters while other bones took over the task of jaw articulation. It would also explain the otherwise peculiar fact that in advanced synapsids, the quadrate was never fused to the skull although it was a crucial part of the jaw support system. An unfused bone, capable of independent vibration, could possibly have better served the task of sound transmission.

In fact, it has recently been suggested (Crompton, 1985) that this selective pressure to increase auditory sensitivity was responsible for what appears to have been three groups of early mammals independently confining the malleus and incus (the articular and quadrate of reptiles) to the middle ear with the already-present stapes: Cretaceous triconodonts; multituberculates; and therian mammals (which include marsupials and placentals).

Among the oldest (Late Triassic) early mammals are the morganucodontids. Of these, the European *Eozostrodon parvus* is known from complete skulls (Figure 6.5) and virtually complete skeletons. While these creatures display more mammalian features than the advanced synapsids (such as a better-developed dentary–squamosal jaw articulation, a relatively larger brain, a wider basicranium, a more extensively developed secondary palate and two types of cheekteeth — molars and premolars), they also carry as phylogenetic baggage a suite of pre-mammal features more characteristic of the synapsids than of later mammals. These primitive features include a distinct articular–quadrate jaw articulation, a relatively posterior and low jaw articulation with the skull, retention of a suite of albeit small postdentary bones and a relatively large stapes.

Clearly, the first mammals to be recognised as such (because of their possession of a well-developed dentary–squamosal jaw joint, loss of alternate tooth replacement of post-canine teeth, modified periotic bones to enclose a larger cochlea, as well as a few other basicranial features, see e.g. Crompton, 1985) are only marginally distinct from the structurally more primitive synapsids. Because of the fossil record, it is clear that the boundary between reptiles and mammals has no significant qualitative borderlines. There is a continuum of fossil forms from basically primitive reptiles, to primitive synapsid reptiles, to advanced mammal-like reptiles, to primitive reptile-like mammals, to advanced mammals of the kind that survive today. The fossil record clearly supports the predictions of the evolutionary model that intermediates between reptiles and mammals would be found.

By the Middle Jurassic (approximately 175 million years ago), the lack of a groove on the inside of the posterior end of the dentary in all groups of mammals except one indicates that the postdentary bones were entirely freed of jaw support responsibilities and presumably were functioning solely as sound transmitters. All three surviving groups of mammals, the monotremes, the marsupials and the placentals, lack postdentary bones attached to the dentary but have four small 'free' bones (Figure 6.8). These 'free' bones are the ectotympanic (which supports the tympanic membrane), the malleus (which is in contact with the ectotympanic and rests against the tympanic membrane and also articulates with the incus), the incus (which articulates with the malleus and the stapes) and the stapes (which articulates with the incus and also rests against the oval window at the entrance to the inner ear). These sound-transmitting middle-ear bones

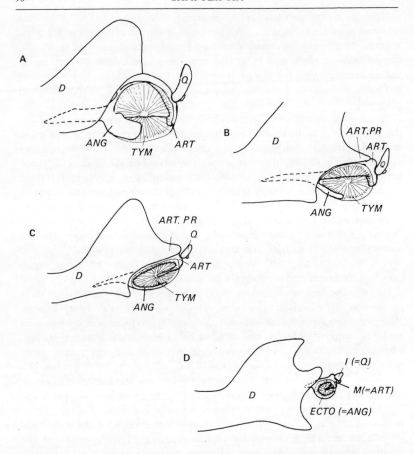

Figure 6.7. The region of the jaw articulation system in: (a) an Early Triassic mammal-like reptile (*Thrinaxodon*); (b) a Middle Triassic mammal-like reptile (e.g. *Probainognathus*); (c) a Late Triassic mammal (*Eozostrodon*); and (d) a living mammal (*Didelphis*). The quadrate (Q) is part of the skull; the other bones have their embryological origin in the lower jaw. The sequence demonstrates the size reduction with time of the postdentary bones and the intimate developmental relationship between the jaw articulation system and the bones of the middle ear. In stages (a)–(c), the postdentary bones were serving two functions: jaw articulation and sound transmission. By the time stage (d) was first achieved (about 174 million years ago) these bones no longer served to articulate the lower jaw (at least in adults — see text for a fuller explanation) but still served to transmit sound. *Abbreviations*: D, dentary; Q, quadrate; TYM, tympanic membrane (the ear drum); ART, articular; ANG, angular; ART.PR, the condyle of the dentary which articulates with the squamosal; M, the mammalian malleus (or hammer); I, the mammalian incus (or anvil); ECTO, the mammalian ectotympanic. (After Allin, 1975.)

are adjacent to the posterior edge of the dentary in exactly the same relationship as the synapsid and primitive mammalian angular, articular, quadrate and incus (Figure 6.8). That they are the same four bones, performing in part the same function of sound transmission, is an inescapable conclusion.

In this regard, studies of the embryological development of the modern mammal jaw articulation system and middle ear are very interesting. Palmer (1913) followed their development in a bandicoot (*Perameles*). When the young are born they are in effect two-week-old embryos. At this stage the young has a functional quadrate–articular jaw joint — a normal reptilian system — and uses it to open its tiny mouth very wide to grasp a teat in the pouch (Figure 6.8c). As the embryonic bandicoot develops in the pouch, the embryonic articular and quadrate become the malleus and incus of the middle ear. The angular matures between the articular and the dentary to become the ectotympanic for support of the developing tympanic membrane. The dentary develops posteriorly and the squamosal bone 'overgrows' the small malleus and incus to make contact with the dentary. By the time (less than sixty days later) that the young bandicoot first releases its grip on the teat it has, like all living mammals (Figure 6.8b) and most fossil mammals since the Middle Jurassic, a fully functional dentary-squamosal jaw articulation system and four 'free' bones in its middle ear.

While it is inappropriate to interpret ancestral adult conditions from the embryological conditions of descendants (Luria *et al.*, 1981), this embryological evidence strongly suggests that there is in living mammals an intimate developmental relationship between the jaw articulation system and the bones of the middle ear. This developmental relationship is precisely mirrored by what one finds in the fossil record of the reptile-to-mammal transition.

To recapitulate, from the fossil record there is clear evidence that the transition from the reptilian to the mammalian jaw articulation system was gradual and that it went bone-in-bone, as it were, with simultaneous changes in the middle ear. The net effect, from the modern mammal point of view, was that the dentary and squamosal replaced the articular and quadrate as the bones involved in articulation of the lower jaw to the skull. The replacement was gradual and involved an intermediate condition, represented by some of the most advanced synapsids as well as by some of the most primitive mammals, in which both sets of bones simultaneously provided support for the lower jaw. Three postdentary bones, the angular and articular of the lower jaw and the quadrate of the upper jaw, functioned simultaneously as transmitters of sound to the stapes which then passed the vibrations on to the inner ear. As the postdentary bones reduced in size and lost their commitment to jaw support, their capacity to transmit high-frequency sounds increased. Subsequent evolution amongst later mammals resulted in eventual confinement of the angular (as the ectotympanic which supports the tympanic membrane), the articular (as the malleus or 'hammer'), the quadrate (as the incus or 'anvil') and the stapes (the 'stirrup') to the middle ear in a position just behind the posterior end of the dentary. The embryological development of the jaw articulation system and the middle-ear region in modern mammals indicates the intimate developmental relationship between the bones which articulate the lower jaw and those which transmit sound, a relationship that is essentially the same as that demonstrated by advanced synapsids.

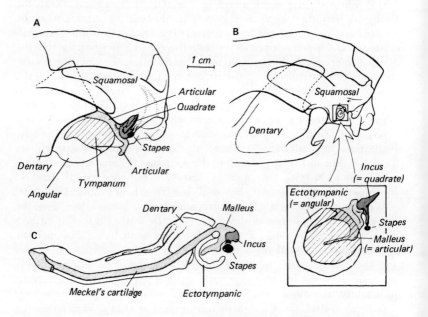

Figure 6.8. This diagram summarises the developmental relationships between the bones of the jaw articulation system in mammal-like reptiles and mammals. In (a) which is an Early Triassic mammal-like reptile (*Thrinaxodon*), the postdentary bones are transmitting sound as well as articulating the jaw. In (b) which is a living mammal (*Didelphis*), the reduced postdentary bones (enlarged in the inset) only transmit sound. In (c) which is a juvenile condition for a living mammal (e.g. the marsupial bandicoots; see text for explanation) the intimate developmental relationship between these sound-transmitting bones and the jaw articulation system are evident. As the animal matures, the squamosal bone of the skull 'overgrows' the ectotympanic (which holds the ear drum), the malleus and the incus which thereby become part of the middle ear, to make contact with the posteriorly growing dentary. The striking similarity of the postdentary bones in the Triassic mammal-like reptile, the relatively unreduced embryological bones which abut the rear of the jaw in modern mammals and the reduced middle-ear bones of the adult mammal into which they develop, provide clear support for the evidence of the fossil record that the mammalian condition developed gradually from that found in the now extinct mammal-like reptiles. (After Crompton and Jenkins, 1979.)

Clearly, the fossil record most definitely provides evidence in support of the existence of former 'links' between what are today significantly different kinds of organisms. In this case, the Triassic synapsids and morganucodontids firmly bridge two otherwise quite distinct classes of vertebrates, the reptiles and the mammals.

The Gap Between Humans and Non-humans

Before the 1850s, it was possible to look around and see nothing but wide gaps between ourselves and all other creatures. In fact, until the discovery of 'Java Man' (*Homo erectus*) in 1893 by Eugene Dubois, there were no significant fossil hominids. The only earlier discovery was a single neanderthal skull (*Homo sapiens neanderthalensis*), dug out of a cave in Germany shortly before Darwin published his *On the Origin of Species*. It was dismissed by an anatomist of the day as the skull of a Cossack soldier who had lost his way while chasing Napoleon's army and perished.

Since then, there have been hundreds of significant discoveries of hominid and pre-hominid fossils in Africa, Europe and Asia. As our knowledge of the fossil record of humans and human-like primates has grown, so too has understanding about the origin and relationships of humans to other primates. The once-yawning gaps have in some cases been completely bridged and in others at least narrowed. Here we will consider only a few aspects of the fossil record and specifically that part of the record that demonstrates how *we* as a species are woven inextricably into the tapestry of primate evolution.

Living primates are represented by seven basically different groups (refer to Figure 6.9), including tarsiers, monkeys, lemurs and humans (the row across the top of the figure). There is still some debate about whether tree shrews (*Tupaia*) should be included among the primates. Figure 6.9 shows 22 of the more than 110 extinct primate genera now known.

The Paleocene *Purgatorius* is a link between the Order Primates and other orders of mammals, specifically the insectivores. Unlike other primates, it retains an insectivore-like third incisor and first premolar. Like later primates, and unlike primitive insectivores, it displays a characteristic enlargement of the central lower incisor and a reduction of the first premolar.

In addition to the many hundreds of extinct primates now known (Szalay and Delson, 1979), linking forms have also been discovered which tie together otherwise distinct lineages of primates (Archer and Aplin, 1984; Gingerich, 1984). Examples of these include *Cantius* (which links the prosimian adapids to omomyids), *Amphipithecus* (which links anthropoids to adapids), and *Aegyptopithecus* (which links pliopithecids to hominoids).

The structural gap between humans and great apes was once considered profound. Fossils are now known which bridge this gap. With the discovery in 1924 of the first skull of *Australopithecus africanus*, it became

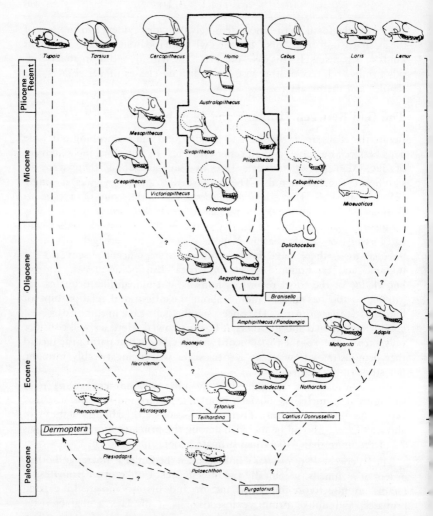

Figure 6.9. Suggested phylogenetic relationships of more than thirty primates. The bottom of the Paleocene represents 65 million years ago. A frame is drawn around the hominoids. See text for further explanation. (After Gingerich, 1984.)

clear that this enigmatic 'ape-man' and its later-discovered relatives represent a group of hominids that sit squarely in the middle of the once profound gap between humans and apes.

Australopithecines display a mingling of ape-like and human-like features. For example, they had: a small brain (450–550 cc; chimpanzees have approximately 400 cc and humans approximately 1400 cc); distinctly bipedal as well as arboreal locomotion; prognathous ape-like jaws; ape-sized teeth but small canines; human-like dental occlusion; (except in the oldest form, *Australopithecus afarensis*) no incisor–canine diastema; and long toes. Other features structurally intermediate between those of apes and humans include: the degree of expansion of the cranial vault; the position of the palate relative to the zygomatic arches; the extent of development of the brow ridges; the shape of the dental arcade (rectangular in apes, U-shaped in australopithecines and parabola-shaped in humans); the relative position of the foramen magnum; the degree of separation of the big toe of the hind foot; and the shape of the foot bones.

The oldest australopithecines are represented by a fragment of a humerus (The 'Kanapoi hominid'), a fragment of a femur and a partial frontal bone. All are about four million years old but too poorly represented to be particularly informative about early hominine evolution. At least three slightly younger species of undoubted *Australopithecus* are much better known. *Australopithecus afarensis*, known from over forty individuals including a partial skeleton, is the oldest of these at about 3.7 to 3.0 million years old. *Australopithecus africanus* occurs in sediments between 2.7 and 2.2 million years old. From one site alone over forty individuals are known including one partial skeleton. The larger species, *A. robustus* occurs in sediments ranging from 2.1 to 1.0 million years old from several areas in Africa.

Discovery in Ethiopia in 1973 of a hominid knee joint and, in the following year, of a dozen additional specimens including a partial skeleton ultimately led to description of *Australopithecus afarensis* (Brace, 1983; Delson, 1985; Johanson and Edey, 1981; Weaver, 1985). The partial skeleton of this species became better known as 'Lucy'. These and other specimens of *A. afarensis* from Tanzania are now the best known of the older hominids. Unlike the younger australopithecines, *A. afarensis* had slightly longer arms and shorter legs, a diastema between the incisors and the canine (as in apes but not as in later australopithecines or hominines), larger canines than occur in later hominids and a more primitive morphology of the first premolar. Otherwise *A. afarensis* resembles the later australopithecines and one of the more primitive species of our own genus, *Homo habilis*.

In a recent symposium on the fossil record of hominoids (Delson, 1985) there was a lot of discussion about *Australopithecus afarensis*. It was generally (but not unanimously) concluded that the fossil material presently referred to as *A. afarensis* did in fact represent a single variable species. Further, it appears on the basis of detailed study of nearly complete skeletons to have been a species that '. . . spent a significant amount of time in the trees without being as adept as an ape and that also lived on the ground without being as quick and agile on two legs as are humans' (Susman *et al.*, 1985: 189). In sum, its features (within single skeletons) are intermediate between those of apes and humans.

Sixty years after the discovery of the first known australopithecines, it is

clear that as a group these hominids include among their ranks our first
cousins and the ancestors of the species of the genus *Homo* (Figure 6.10).
While they display various human-like features not exhibited by apes, they
also retain a large number of ape-like features that were present in the
common ancestor of the human–australopithecine line on the one hand and
the lineage that led to modern apes on the other. The earliest australopithecines demonstrate a condition structurally intermediate between apes and
Homo. Although they stood upright, had small canines and other 'human'
features, they had very small brains (450–530 cc), well-developed faces and
other features which are more ape-like. *Australopithecus afarensis* is the oldest

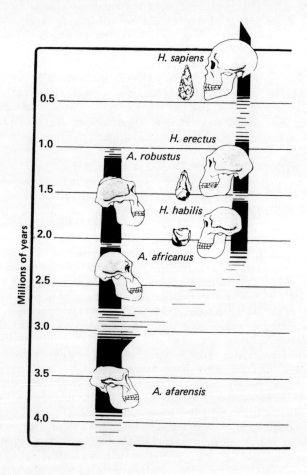

Figure 6.10.   Time range and structural sequence of species of *Australopithecus* and
Homo. (From Johanson and Edey, 1981.)

known hominine, and is slightly more primitive than *A. africanus* which is, in turn, older and more primitive than the oldest known species of *Homo* — as well as another of its descendants, *Australopithecus robustus* — and so on up the sequence. *Homo habilis* (about 1.5 to 1.8 million years old) is a perfect mixture of *Australopithecus* and *Homo* features. As a result it is sometimes placed in the genus *Australopithecus* rather than in *Homo*. Whatever genus it is placed in, it is an excellent link between the two genera. Transitional populations and/or specimens are known between all taxa shown in Figure 6.10 and it has become difficult to say where in time one species ends and its descendant starts. The first evidence of manufacture of stone tools is associated with *H. habilis*. Fire was first used by *H. erectus* and agriculture and atom bombs by *H. sapiens*.

It is now difficult to rely on breaks in the fossil record to provide temporal boundaries between species. Starting from the oldest known forms, *H. habilis* is about halfway between australopithecines and *Homo* and an excellent link between the genera. The best-known specimen is Richard Leakey's 'ER 1470' skull from northern Kenya (Leakey, 1981; Weaver, 1985). It displays an australopithecine-like face and molar alveoli but different tooth proportions. Like species of *Homo* but unlike species of *Australopithecus*, it has proportionately smaller cheekteeth relative to the size of the anterior dentition. It also has a brain capacity of 750+ cc which is halfway between species of *Australopithecus* and *Homo*. The mix of features is in fact so striking that some anthropologists have recently referred to this taxon as *Australopithecus habilis* (Brace, 1983). Appropriate nomenclature, however, is less important here than recognition of the structural and temporal continuity between the two otherwise distinct lineages of hominids.

The next youngest species of our genus is the much more widespread *Homo erectus*. The transition from *H. habilis* to *H. erectus* involved a steady increase in brain size from about 800 cc in the early *erectus* populations to 1200 cc in the later populations (Wolpoff, 1984).

Homo erectus occurs in sediments ranging in age from 1.5 to about 0.3 million years old. It was the first species of our genus to leave Africa, the first hominid to make use of fire and the first to make clothes.

Populations of *H. erectus* from China have been known as Peking Man, those from Indonesia as Java Man and those from Europe as Heidelberg Man. With the more recent growth in knowledge of *erectus*-type material from many additional fossil localities in Africa and elsewhere, a few anthropologists (e.g. Stringer, 1984) have suggested that we should recognise more than one species within what is presently called *H. erectus*. However, although there is debate about whether or not more than one species should be recognised among this material, there is agreement that all of these *erectus*-like populations represent the same grade of hominid, one structurally and temporally intermediate between the older *H. habilis* and the younger *H. sapiens*.

In terms of our interest in the capacity of the fossil record to demonstrate links, *H. erectus* is of particular interest because it was clearly the one that eventually gave rise to our own species, *H. sapiens*. A recent study of all known *H. erectus* specimens demonstrates a continuous change in morphology throughout the time span of the species from the more 'classic' *erectus*-type features of the older African (Turkana) and Indonesian (Sangiran) populations to the more *sapiens*-like features of the populations from Ngandong (Wolpoff, 1984).

The overall transition from *erectus*-type morphologies to *sapiens*-type morphologies is so well documented that some populations intermediate in morphology and age simply cannot be placed definitely in either species. For example, the skulls from deposits along the Solo river at Ngandong, Java, estimated to be about 120 000 years old, represent perfectly intermediate morphologies and appear to represent populations living at the time when the *erectus–sapiens* transition was taking place (Brace, 1983).

Some anthropologists who have studied the *H. erectus* and early *H. sapiens* material suggest that because it demonstrates such a gradual transition, there is no point in even trying to distinguish *H. erectus* from *H. sapiens*. They suggest that we should regard them simply as ancestral and descendent phases of one species, *H. sapiens*. Other anthropologists (e.g. Rightmire, 1985) who have looked at this transitional sequence, suggest that there are a few features that can be used to separate the sequence into two forms. However, all anthropologists who have recently studied this material agree that there can be no other conclusion than that *H. sapiens* developed from *H. erectus*, whatever taxonomic distinction is given to the *H. erectus* material.

Because *erectus*-type morphology grades into *sapiens*-type morphology, it is difficult to say precisely when *H. sapiens* made its first appearance in the fossil record. Among the structurally intermediate populations whose taxonomic position is unclear there are many 'archaic' types of *H. sapiens*, some as old as about 0.3 million years. Less controversial early *sapiens* material ranges in age from about 75 000 to 150 000 years. This includes skulls from Broken Hill (Africa) that are about 150 000 years old and skulls from Petralona (Europe) 200 000 and Laetoli (Africa) approximately 120 000 years old. Neanderthals lived during the last glacial period about 35 000 to 70 000 years ago and evidently represented a regionally distinct population of *H. sapiens*.

A review of the fossil record for the times of appearance of *human* features demonstrates that the majority of these most clearly did not develop in our species but rather were a cumulating legacy passed on from our many distant and often non-human ancestors. Without the evidence of the fossil record, these features would be unknown in any creatures apart from humans.

The following is a summary of these features detailed in part by Badgley (1984). Thick enamel on the teeth first appeared about seventeen million years ago. Bipedalism was present in the oldest known *Australopithecus* at least as early as four million years ago. The short and wide ilia of the pelvis as well as other postcranial features associated with an upright stance were present at least as early as four million years ago (in *A. afarensis*). Small canines, parabola-shaped tooth rows and double-cusped first lower premolars were developed by the early australopithecines as early as three million years ago. *Australopithecus afarensis* had only just begun to develop a double-cusped lower first premolar. Stone tools were manufactured at least as early as 2.1 million years ago, *Homo habilis* presumably being the maker since no australopithecines have been shown to manufacture stone tools. Speech may have been in use as early as 1.8 million years ago in *H. habilis*. Rapid expansion of the brain began about three million years ago in the australopithecines but the brain achieved modern size approximately one million years ago in advanced *Homo erectus*. Religious belief in an after-life was present among neanderthals (*Homo sapiens*) possibly as early as 70 000 years ago. Agriculture began to be practised about 10 000 years ago by *H. sapiens*.

Before leaving the primates, we should consider one other facet of interest. The link between humans and apes indicated by the extinct australopithecines has been independently confirmed by studies of molecular biology. It has been suggested (Badgley, 1984) that because comparisons of nucleic acids and proteins involve studies of genes and immediate gene products, they are more reliable indicators of relationship than are studies of tooth or skeletal morphology. Whether or not this is correct, the pattern of relationship indicated by molecular studies (Figure 6.11) basically conforms with that indicated by morphological studies of living and fossil primates.

Figure 6.11 is a diagram that summarises knowledge about the relationships of the 'higher' primates discussed here. This sort of diagram is called a cladogram. In it, relationships are indicated by the branching scheme. All branches that share a common fork represent lineages that are more closely related to each other than they are to any lineage that forks below that point. For example, the species of *Australopithecus* are more closely related to the species of *Homo* than they are to the chimpanzees because they share with species of *Homo* certain specialised (derived) features (such as tooth shape, bipedalism etc.; see above) which are lacking in the chimpanzees. Similarly, the species of *Proconsul* are more closely related to the species of *Homo* than they are to the species of *Aegyptopithecus* because the latter lacks derived features shared by *Proconsul* and all of the other primates placed to the right of *Proconsul*. In the same way, the species of *Sivapithecus* as depicted here are more closely related to those of the orangutan than either is to the species of *Ramapithecus*, although all three are more closely related (through a common ancestor) to the hominines than they are to gibbons or any of the other groups to the left of the *Ramapithecus–Sivapithecus*–orangutan branch.

The numbers placed along the 'main stem' of the cladogram represent the approximate time (in millions of years) when the branch indicated separated from branches to the left of it. These figures represent either radiometric dates (e.g. those of 3 and 25 million years old (myo)) or dates based on the 'molecular clock' which has been determined on the basis of genetic difference that separates living species on the branch indicated from those on the branch to the left of it (e.g. 5 and 18 myo).

The oldest known occurrences of extinct forms (indicated by a dagger) are given in parentheses below the names. For example, the oldest known fossils referable to *Australopithecus afarensis* are approximately 3.7 million years old.

The oldest and most primitive hominoid is *Aegyptopithecus zeuxis* from the middle Oligocene of Africa. The youngest hominoids include the living apes and humans. The most primitive living apes appear to be the gibbons, followed by the orangutan and its ancestral relative *Sivapithecus*. Amongst apes, chimpanzees show greatest biochemical similarity to humans, differing in only 0.8% of forty blood proteins studied. Tiny differences of this sort are of the kind which, in other types of organisms, indicate species-level distinctions within a single genus. The 'molecular clock' date for

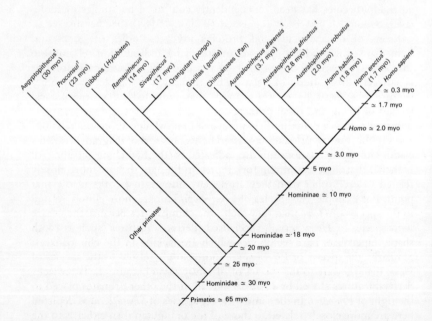

Figure 6.11. Diagram of relationships of members of the superfamily Hominoidea. See explanation in text. Data supporting the diagram are morphological, palaeontological and molecular. The dagger indicates extinct forms. Drawing by J. Taylor.

separation of chimpanzees and humans indicates a shared common ancestor no more than six million years ago. Even closer to humans, however, are the australopithecines.

In addition to biochemical similarity, modern humans share with chimpanzees and gorillas many striking morphological features lacking in other great apes including (Groves, 1985): axillary organs (clusters of apocrine sweat glands in the armpit); many modifications of the hind legs and feet such as long and broad heels; a thickly haired scalp; ear lobes; frontal sinuses in the skull; delayed dental eruption; loss of the os centrale; and reduced baculum (bone in the penis). Similarly, humans, chimpanzees, apes and the orangutan share features that are lacking in other living great apes (the gibbons) such as: a less hairy skin-covering (gibbons have 700 or more hairs per cm^2 in contrast to the other great apes and humans which have at most 200 hairs per cm^2 — humans appear to have much less hair than chimpanzees but this appearance is mainly due to the fact that human hair is shorter and less pigmented); and a more flexible wrist (Groves, 1985).

In this way, it is easy to see how, using just morphological features and molecular similarities of living primates, a pattern of relationship emerges that ties humans unambiguously to apes. Further, it is clear that this pattern arises independently from the basically similar one which has emerged from study of the fossil record.

Although problems of classification are not a focus here, it is interesting to note that there is no support in this pattern of relationships for recognition of humans as belonging to one family (Hominidae) and apes to another (Pongidae). Clearly, in terms of molecular as well as morphological features, humans are closer to chimpanzees and gorillas than chimpanzees and gorillas are to orangutans. In other words, despite our well-developed sense of self-importance, we are at base simply another kind of ape. Failure to recognise this close relationship can only be defended by ignoring the biological reality of the situation.

SUMMARY

There is ample evidence from the fossil record for structural links between what otherwise appear to be fundamentally distinctive types of living organisms. We have considered only three examples of this, the links between birds and reptiles, between mammals and reptiles and between humans and other primates. However, there are many other examples of significant links found in the fossil record such as forms intermediate between whales and ungulates (e.g. the Early Eocene protocetid *Pakicetus*; see Gingerich, 1983), horses and primitive condylarths (e.g. the Paleocene phenacodontids) and even invertebrates and vertebrates (e.g. Cambrian chordates such as *Pikaia*). The point is that the concepts of evolution have received only support from the fossil record.

FURTHER READING

Anon. (1980) *Man's place in evolution*. British Museum (Natural History), London.

Broadhead, T.W. (ed.) (1984) *Mammals*. Notes for a short course organised by P.D. Gingerich and C.E. Badgley. University of Tennessee, Department of Geological Sciences Studies in Geology 8: Knoxville

Cherfas, J. (ed.) (1982) *Darwin up to date*. IPC Magazines Ltd (as 'A New Scientist Guide'), London.

Crompton, A.W. (1972) The evolution of the jaw articulation of cynodonts. *In* Joysey, K.A. & Kemp, T.S. (eds) *Studies in vertebrate evolution*. Oliver & Boyd, Edinburgh: 231–251.

Crompton, A.W. & Jenkins, F.A. (1979) Origin of mammals. *In* Lillegraven, J.A., Kielan-Jaworowska, Z. & Clemens, W.A. (eds) *Mesozoic mammals: the first two thirds of mammalian history*. University of California Press, Berkeley: 7–58.

Gingerich, P.D. (1983) Evidence for evolution from the vertebrate fossil record. *Journal of Geological Education* 31: 140–144.

Gould, S.J. (1977) *Ontogeny and phylogeny*. Belknap Press of Harvard University Press, Cambridge, Mass. 501 pp.

Gould, S.J. & Eldredge, N. (1977) Punctuated equilibria: the tempo and mode of evolution reconsidered. *Paleobiology* 3: 115–151.

Kemp, T.S. (1982) The reptiles that became mammals. *In* Cherfas, J. (ed.) *Darwin up to date*. IPC Magazines, London: 31–34.

Patterson, C. (1978) *Evolution*. British Museum (Natural History),London & the University of Queensland, Brisbane. 197 pp.

Raup, D.M. & Stanley, S.M. (1978) *Principles of paleontology*. W.H. Freeman, San Francisco. 481 pp.

Rensberger, B. (1984) Bones of our ancestors. *Science 84* 5(3): 28–39.

Rose, K.D. & Bown, T.M. (1984) Early Eocene *Pelycodus jarrovii* (Primates: Adapidae) from Wyoming: phylogenetic and biostratigraphic implications. *Journal of Paleontology* 58: 1532–1535.

Smith, J.M. (ed.) (1982) *Evolution now: a century after Darwin*. *Nature* and Macmillan, London. 239 pp.

Stringer, C. (1982) The evolution of man. *In* Cherfas, J. (ed.) *Darwin up to date*. IPC Magazines, London.

Washburn, S.L. (1978) The evolution of man. *Scientific American* 239: 146–154.

Wood, B.G. (1976) *The evolution of early man*. Eurobook, London.

Zihlman, A.L. (1982) *The human evolution coloring book*. Barnes & Noble Books, New York.

CHAPTER SEVEN

SQUARING OFF AGAINST EVOLUTION: THE CREATIONIST CHALLENGE

Michael Archer

> It is crucial for creationists that they convince their audience that evolution is not scientific, because both sides agree that creationism is not. (Miller, 1982: 4).

INTRODUCTION

In earlier chapters I have given an introduction to evidence supporting the concepts of evolution provided by living organisms and the fossil record. There is overwhelming scientific support for the concepts of evolution. Like apples falling from trees, evolutionary change is a fact, one that has been witnessed in the wild and in laboratories as well as evidenced by the fossil record.

As in any science, however, concepts of evolution must always remain vulnerable to falsification and there must never be off-handed dismissal of any claim that the theory of evolution has been falsified. Fundamentalist Christians who perceive the theory of evolution as a threat to their religious beliefs and to the notion that mankind is God's special Creation have striven valiantly to falsify every aspect of the evolutionary model that conflicts with a literal interpretation of the Bible. These, and any other attempts at falsification, when honestly conducted, are perfectly justifiable and in the long run essential to establish whether there is any reason to abandon the evolutionary model in favour of some other.

At first, fundamentalist Christians did not attempt to conceal the fact that the sole basis for their challenge to evolution was their belief in the infallibility of the Bible. Such a belief, of course, is not a valid test of the evolutionary model because it depends on *a priori* acceptance of an alternative model: a literal interpretation of Genesis. Creationism, to the extent that it makes assumptions about the literal truth of the Bible, is clearly religion, not science.

This honest admission by Creationists was the main reason for their lack of success in forcing the teaching of Creationism in science classes.

Faced with repeated failures in their efforts to get Creationism into the school system while being honest about their beliefs, they resorted to deceit. Under the guise of 'scientific Creationism', they have attempted to obscure the fact that, at base, Creationism is a religious discipline, not a scientific one. As has been pointed out elsewhere, to become a member of the Creation Research Institute applicants must have not only a degree but also the religious commitment to sign a document to the effect that they acknowledge the Genesis account of the Bible as the literal truth about the origin of all basic types of living things, including man. This commitment simply means that, as Creation 'Scientists', they cannot accept any evidence that contradicts the Genesis account of Creation. As a result, they cannot use the scientific method to examine questions about Creation because they have, *a priori*, denied themselves the opportunity to consider any other than supernatural explanations.

However, we should set this intellectual incapacity aside because, although it indicates why Creation 'Scientists' cannot do real scientific research in this area, it does not answer the more crucial question about whether the Creation 'Science' model can be falsified and, if so, whether it has survived attempts at falsification.

Duane Gish, one of the most outspoken Creation 'Scientists', clearly considers that Creation 'Science' cannot be falsified: 'Creation is, of course, unproven and unprovable by the methods of experimental science. Neither can it qualify, according to the above criteria, as a scientific theory, since creation would have been unobservable and would as a theory be nonfalsifiable' (Gish, 1979: 21–22). He goes on to say that this should not worry Creation 'Scientists' because the evolutionary model is similarly nonfalsifiable.

Gish is wrong on both accounts. Predictions arise from the Creation 'Science' model that can be and most certainly have been tested (Ruse, 1982). Gish himself (1979: 49) states a condition that would falsify the model: 'As a matter of fact, the discovery of only five or six of the transitional forms scattered through time would be sufficient to document evolution'. Some of the predictions that arise from the Creation model and which contrast with those arising from the evolutionary model are considered below.

In the same way, predictions arising from the evolutionary model can be tested and the outcomes of some of these tests have been discussed in earlier chapters. Further, it should be noted here that many of these predictions (and in contrast to those arising from the Creation model) can be tested experimentally. Examples of these include predictions about the existence and effects of natural selection. About the inability of the Creation model to be tested *experimentally*, Gish is trivially correct. But Gish's statement that the model cannot be tested in any way is nonsense.

In their attempts to defend their convictions about the origin of life on Earth without using as up-front evidence their faith in the infallibility of the Bible, Creation 'Scientists' have attempted to demonstrate that observations which have led scientists to conclude that the natural world provides evidence for evolution in fact fail to do so. In casting doubt on the whole concept of evolution, they evidently intend to leave the disillusioned with no alternative but to embrace their own understanding of Biblical Creation.

Creation 'Scientists' have made claims that the evolutionary model has been falsified. If these are correct, evolutionists would have to discard or revise the basic evolutionary model. What are these claims for falsification?

According to Creation 'Scientists', evolution (as presently conceived by scientists) is impossible. This is so for essentially three reasons: (1) evolution as a natural process would violate the strictures of the second law of thermodynamics; (2) evolution of life from non-life and the evolution of complex life forms from simple ones violates all laws of probability; and (3) even if evolutionary processes *did* exist, the Earth is much too young to have enabled evolution to produce the variety of known forms of life.

Creation 'Scientists' have also challenged the evidence for evolution from the fossil record. Their basic challenges are: (1) that there are no links between the different kinds of organisms; (2) that there is evidence for the coexistence of all kinds of organisms from trilobites to dinosaurs to humans (as, coincidentally, is suggested in the Genesis account of Creation); and (3) that there is evidence that all kinds of life on Earth were drowned together (except for a boatful which were saved by Noah) in a worldwide flood which occurred sometime this side of 20 000 years ago.

CHALLENGES TO THE PLAUSIBILITY OF EVOLUTION AS A WHOLE

Evolution of life as a violation of the second law of thermodynamics

The second law of thermodynamics states that 'The entropy of an isolated system can never decrease' (Freske, 1981). Entropy may be defined as a measure of disorder or reduced energy state. In biological terms, the growth of a mouse, for example, would lead to an overall decrease in disorder of the materials of which it is comprised and hence an overall decrease in entropy. Death of the mouse would be an increase in its disorder and hence an increase in entropy. The law implies that isolated systems naturally tend to change from ordered to less ordered and in so doing tend towards increased entropy. By 'isolated system' is meant one that cannot exchange energy or matter with its surroundings. With respect to the relevance of this law to the possibility of evolution, Creation 'Scientists' interpret the law to mean that life cannot, of its own accord, become more highly organised (i.e. evolve) because this would result in a decrease in entropy.

The problem for Creation 'Scientists' is that the second law applies only to systems that are *isolated*; that is, systems that acquire no more energy than they started with. Life is *not* such a system. Energy constantly pours into the Earth in the form of radiation from the sun where chemical energy is converted (in a 'downhill' fashion towards greater entropy) into radiant energy. Plants convert simple molecules into more complex ones using this energy and animals build themselves into complex organisms by utilising the energy from plants or other animals.

If the capacity for living systems to reduce entropy and grow were not possible, the seeds of trees would not be able, by organising elements and compounds around them, to develop into adult trees and every human being conceived would shrivel as a tiny zygote. The process is certainly inefficient and there is much wastage but solar energy continues to pour in. Life is, in this sense, a parasitic eddy adjacent to and dependent on the sun, whose energy state as a whole is degrading. It is rather like a backward-flowing eddy which develops adjacent to and is driven by the main flow of a stream. The net change in the whole isolated system (the whole of the stream) is indeed increased entropy (the water flows downhill and loses more of its potential energy) but along the way this same flow continually supports smaller flows in the opposite direction, that is towards decreased entropy. There is quite clearly nothing impossible about this and it in no way contradicts the second law.

Furthermore, just as individuals can develop from single cells to complex structures without violating the second law of thermodynamics in the process, so too can simpler forms of life develop into more complex ones. The way in which this happens is again dependent on the availability of an energy flow. The simple mechanism for increasing complexity is the overproduction of non-identical young. More young are produced than can survive and natural selection 'tosses away' less fit individuals. The individuals whose genetic constitution better suits them to survive will pass the variation to the next generation. The pool of variation for natural selection to work with steadily changes through mutation and genetic recombination. While most individuals will have unsuitable genetic instructions and be selected out, a few individuals will have successful instructions and survive to reproduce. Because the environments in which life occurs are constantly changing, there will be a net change in life forms through time as species are moulded by natural selection to track those changes. These developments are again dependent on a supply of energy.

The net 'cost' of this steady increase in complexity is high and can be measured as the energy value of all individuals which fail to reproduce combined with the energy value of the metabolic wastes and heat dissipated by all of the individuals produced (discards and survivors alike). The total reduction in available energy represented by these losses (i.e. the net increase in entropy) vastly exceeds the decrease in entropy represented by the survivors. There is, therefore, despite the continuous input of energy, a constant increase in entropy. The net increase in entropy is the 'cost', unwittingly paid, that enables life to run counter to the main current and steadily increase in complexity with time.

In other words, the capacity of living organisms to persist, become better adapted and more complex with time depends upon a constant input of energy. The energy source is ultimately the 'downhill' flow of energy from the sun. The net effect can be, and has been, a progressive increase in complexity of life forms without any violation of the second law of thermodynamics. Creation 'Scientists' who fail to recognise that this is possible do not, cannot or will not understand the second law of thermodynamics. Evolution of life has not occurred in a closed system and hence presents no contradiction to the law (Freske, 1981; Thwaites and Awbrey, 1981; Patterson, 1983).

The improbability of evolution

Considering improbability arguments in general, Creation 'Scientists' frequently use these to ridicule the possibility of the origin of life and its subsequent evolution. For example, an often-used Creationist analogy is that saying life could have originated spontaneously is similar to suggesting that a junkyard could spontaneously convert itself into a jet airliner. Similarly, they rhetorically ask how improbable would it be for a watch to form naturally from a random collection of atoms and, if a single cell is much more complicated than a watch, how less likely would it be for the cell to develop spontaneously?

In fact, Creation 'Scientists'' arguments about the improbability of evolution are seriously misleading. In large part, this is because analogies such as these are totally inappropriate. No serious evolutionist suggests that a cell *did* develop all at once. Rather, they hypothesise that the first cells were the end-product of a long series of evolutionary events which individually are not only probable, but in many cases have already been duplicated in laboratory experiments (see below).

Further, Creation 'Science' arguments are often based on the assumption that an end-product (be it jetliner or protein molecule) is the only end-product which need concern the improbability argument. But this is also grossly misleading. It is well known, for example, that the precise sequence of all amino acids in a protein is not critical — it is the *function* of that protein that matters. In this regard, it is abundantly clear that there can be strikingly different sequences in proteins that will nevertheless perform the same function (Doolittle, 1983). An example of this noted earlier (Chapter 3) is the more than 200 different varieties of haemoglobin molecules in just humans alone.

It is quite clearly the case that many different sequences can be 'acceptable' outcomes in the lottery of evolving organic molecules. To ignore this important fact seriously impairs the whole basis of these improbability arguments. This, incidentally, is one of four fatal flaws woven into Yockey's (1977) widely cited arguments for the improbability of evolution of the cytochrome C molecule (Doolittle, 1983).

The other general aspect of molecular evolution which Creation 'Scientists' notoriously forget about when they use these improbability arguments is the nature of organic molecules themselves. These have inherent associative and aggregative properties (see below) which markedly increase the probability that self-replicating molecules and simple cells would quickly evolve once basic amino acids and nucleotides had developed.

In the same way, snowflakes form spontaneously and naturally because attributes inherent in water molecules predispose them to undergo this striking shape and textural transformation. No divine intervention is necessary to convert the simple water molecule into one of the most beautiful and complex objects known. Life, in this sense, is one of the possible snowflakes of spontaneously formed organic molecules.

Creation 'Scientists' focus on three aspects of evolution which they declare too improbable to have occurred: (1) the improbability of a natural origin of life; (2) the improbability of 'progressive' evolution because of the negligible occurrence of favourable mutations; and (3) the improbability of links because of the non-functional condition of intermediate forms. Let us consider them in order.

There are several areas of research presently underway which demonstrate that the spontaneous evolution of simple life on Earth is far from an improbable event. In fact, one of the most striking aspects of the fossil record is how quickly life *did* appear once suitable conditions had developed. The age of the oldest known sedimentary rocks is about 3 700 000 000 years and the age of the first fossils (simple prokaryotic cells) about 3 400 000 000 years. For obvious reasons it is difficult to determine from the fossil record exactly how the first life developed but experiments have been very instructive about probable pathways (Dickerson, 1978; Doolittle, 1983).

Dickerson (1978) divides the hypothesised transition from non-living matter of the Earth to living cells into five steps: (1) the formation of the Earth with a mixture of gases in the atmosphere that could form the raw materials of life; (2) the synthesis of these materials into biological monomers such as amino acids, sugars and organic bases; (3) polymerisation in water of these monomers into primitive proteins and nucleic acid chains; (4) isolation of droplets of an aqueous solution of these polymers and other materials into independent protocells; and (5) development of some type of reproductive system to enable the protocells to perpetuate themselves.

The first step would have occurred naturally as an outcome of the cooling of the primitive planet. The primordial atmosphere probably contained large amounts of water vapour (from volcanic outgassing), methane, hydrogen, carbon dioxide, carbon monoxide, ammonia, nitrogen and hydrogen sulphide but very little oxygen. Oxygen eventually became a significant component of the atmosphere as the result of metabolic activity of organisms with oxygen-producing photosynthetic systems (probably blue-green algae) which, once established, changed the atmosphere and destroyed the conditions essential for the initial production of life. Free oxygen would have oxidised the chemicals needed for the production of life before they could have become involved in the evolution of protocells.

The other steps postulated by Dickerson have been or are now being tested in laboratory experiments. Amino acids, sugars, and nucleotide bases, which are the major components of the genetic material of modern life, have all formed in experiments which simulate the prebiotic conditions of Earth (Orgel, 1982; Figure 7.1). In early experiments carried out by Stanley Miller and Harold Urey (using just methane, ammonia and water through which were passed electric discharges) quantities of several amino acids were produced including glycine, alanine and aspartic acid. In similar experiments involving other primordially — available materials including nitrogen and hydrogen, many additional amino acids such as leucine, isoleucine, serine, threonine, asparagine, glutamic acid, aspartic acid, adenine, pyrimidine, lysine, phenylalanine and tyrosine have been formed.

Similarly, polymers have spontaneously appeared in laboratory experiments simulating primordial conditions in which solutions of the monomers could have dehydrated along the margins of an ancient sea. Some of the most interesting of these are the studies of Fox (1981, 1984) and Fox and Matsuno (1983) in which it was shown that dry mixtures of amino acids will polymerise spontaneously at temperatures as low as 130°C to produce what Fox calls thermal proteinoids. Depending on other chemicals present, polymerisation will even occur spontaneously after a day or so at 60°C. Some polymers so produced consisted of chains of 200 or more amino acids.

Fox has also reported a most interesting attribute of his spontaneously — produced thermal proteinoids. When they are heated in concentrated aqueous solution to between 130 and 180°C, they transform into microspheres one or two micrometres in diameter. These microspheres develop an outer boundary that resembles a double-layered cell membrane (though it does not contain lipids). If the conditions are right, the microspheres grow, bud and divide in a singularly bacterium-like manner.

Dickerson (1978) demonstrated how DNA, one of the most vigorous self-polymerising molecules, may well have developed from nucleic acids which formed part of protocells such as these. Doolittle (1983) has pointed out that nucleotides (the molecular 'bricks' that comprise DNA and RNA) have in simulation experiments spontaneously polymerised into polynucleotide chains involving as many as twelve nucleotides (Fakhrai et al.,

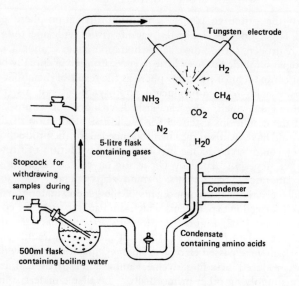

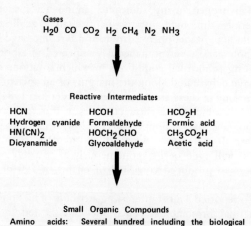

Figure 7.1. Experiments simulating the formation of molecules in Earth's primordial atmosphere. Electric sparks mimicking lightning discharges were passed through a sealed vial with various mixtures of gases representing the primitive atmosphere. Simple amino acids and more complex compounds were produced, depending on the conditions.

1981). Furthermore, recent research by Orgel, Eigen and others has demonstrated that these small chains of nucleotides often proceed to self-replication. When these self-enlarging organic molecules get to be twenty to twenty-five amino acids in length, they can spontaneously double their length (Doolittle, 1983; Miller 1982).

What is more, it has even been found that by a natural process of 'bootstrapping', two strands of nucleotides will partially overlap if their terminal sections comprise complementary sequences. At this point, all along the lengths of the overhanging ends, the nucleotides will spontaneously assemble the missing parts to finish the complementary strand to which it became partially joined. If these paired strands are exposed to temperatures of 80–90°C, they will separate. Once separated, they will again proceed to build complementary strands which faithfully (apart from occasional accidents — i.e. mutations) reproduce the nucleotide sequence that had been formed by the 'bootstrapping' process (Doolittle, 1983).

While the experiments of Fox and others have not so far produced life precisely as we know it, they demonstrate that inherent in prebiotic molecules are chemical properties that can lead spontaneously to life-like attributes. Since the microspheres survive by consuming resources in their environment and are capable of division it is not difficult to see how natural selection could, with the whole Earth as a laboratory and 200 000 000 years of experimentation time up its sleeve, begin to mould such microspheres into simple prokaryotic cells.

The supposed improbability of 'progressive' evolution once life had appeared is even more easily dealt with. It depends in the first instance on the observation that most mutations that affect an organism's structure are either of no value or harmful to its survival. Very few are beneficial. How then, Creation 'Scientists' ask, can random mutations be responsible for the production of complex life forms? To use an analogy, they point out the improbability that a monkey placed in front of a typewriter would ever produce a Shakespearian sonnet by a simple chance striking of keys.

The problem with the analogy is that it lacks one vital part: an automatic censoring system that obliterates every 'mistake' the monkey makes and permits only 'correct' letters, struck by accident, to remain. This automatic censoring system is analogous to natural selection. Mistakes are not permitted to survive if they are disadvantageous to the organism in which they occur but beneficial changes, no matter how rare, are preserved. With this addition to the monkey and typewriter analogy, the probability of the monkey producing a Shakespearian sonnet, if that is the only acceptable end-product, approaches certainty.

While Creation 'Scientists' are correct in saying that most mutations are neutral or deleterious, they are wrong when they extend the observation into a declaration that therefore life could not, by means of chance mutations, have evolved in a progressive manner. Even if only one in a million mutations were favourable, the evolutionary model predicts that such rare events would be preserved and accumulated by means of natural selection and that deleterious ones would be discarded. The net effect of such a process could only be 'progressive' evolution.

There have been many demonstrations of the appearance of new mutations which have resulted in improved fitness of the individuals which possess them (see Further Reading). Betz *et al.* (1974) demonstrated the spontaneous appearance in bacteria of mutant enzymes that enabled utilisation of amides that could not be utilised by the parental stock. A review of many similar examples of favourable mutations is given by Mortlock (1982). Along similar lines, Ohno (1984) predicted how unique proteins could be produced by an organism's use of a mutant coding sequence in the DNA. Because other non-mutant copies of the nucleotide sequence existed as well, both the new and the old enzymes could exist together. Ohno demonstrated, by analysis of a 1295-base-long segment of DNA of laboratory populations of *Flavobacterium*, that this is what appears to have enabled laboratory strains of the organism to degrade nylon by-products, something which the parental strains were not able to do.

The third argument about probability concerns Creation 'Scientists'' difficulties in perceiving how, if evolutionary change occurs gradually, intermediate conditions could persist which fell short of the conditions we see today. How, for example, could an incomplete eye have been preserved through natural selection in order to allow evolution to complete the transition to a complete eye? Similarly, how could complex biochemical pathways such as those of photosynthesis evolve through intermediate conditions when the whole pathway has to exist for the end result to be achieved?

There are a number of solutions for this apparent dilemma. In the case of structural features such as eyes, hands, wings and so forth, the simplest explanation is that there is no logical reason to assume that intermediate structural conditions could not have had adaptive value. For example, take the case of eyes. Almost every conceivable degree of development of eyes, ranging from the light-sensitive eye-spots of single-celled protistans to the complex eyes of vertebrates, can be found as functionally adaptive structures in living creatures. A study by Salvini-Plawen and Mayr (1977) has shown that photoreceptors have evolved independently at least forty times and that in '...more than 20 phyletic lines a series of ever more perfect eyes among still-living relatives can be constructed.... This includes at least 15 lines which — by convergence — have evolved eyes with a distinct lens.'. It is not difficult to see how any refinement in a system which increases sensitivity to the immediate environment would benefit its owner and result in positive natural selection.

Some Creation 'Scientists' ask 'Why, then, if the more complex eye of a hawk for example evolved gradually as the result of constant selective pressure in vertebrates to improve visual sensitivity, haven't garden snails (gastropod molluscs) also evolved similarly complex eyes? Presumably the same selective pressures should have been working on them.'. The obvious answer is that there is no reason to presume that precisely similar selection pressures have been at work on garden snails. They have very different life styles from hawks, live in very different environments and presumably face very different selection pressures. If, as seems probable, there is no selection pressure acting on garden snails to develop eyes with the long-distance visual acuity of hawks, why should they have developed eyes any more complex than their way of life requires?

That is not to say, of course, that molluscs living in environments where greater visual sensitivity would be an advantage could not develop, via gradual evolution, more complex eyes. One has only to consider the relatively much more complex eyes of the fast-swimming predatory molluscan octopus to understand that, if there is a selective pressure to develop greater visual acuity, even such 'lowly' things as molluscs can, through natural selection and evolution, rise to the occasion. Even amongst gastropods, the group of molluscs which contains snails, eye development ranges from relatively simple photosensitive eye pits (e.g. species of *Patella*) to enclosed but simple eyes (e.g. species of *Turbo*) to relatively complex eyes, some even with lenses (e.g. species of *Murex* and *Nucella*) (Salvini-Plawen and Mayr, 1977). Clearly, there is no problem with the idea that structurally intermediate conditions could be functional. Graphic demonstrations of functional intermediate conditions surround us.

Considering further this 'problem' about the possibility of evolving complex structures which would involve supposedly non-functional intermediate stages, there is another evolutionary concept relevant here called preadaptation. This concept implies that natural selection acts on already existing variation but does not cause it. In other words, a population can respond to a selective pressure because of genetic variation already existing in the population. Consider the possibility that the 'ultimate' utility of a structure to an organism may not be the same as the original value of a less complex intermediate condition. An example of this may be feathers. Apart from facilitating flight, there are many possible uses for feathers which would not require them to be of a size, shape and disposition to enable flight. Embryologically, feathers are modified reptilian scales and the fossil record demonstrates primitive feathers, as in *Praeornis sharovii*, structurally intermediate between the scales of reptiles and the feathers of *Archaeopteryx lithographica*, the first known bird. The feathers of *Praeornis sharovii* would not have enabled its owner to fly but, as Ostrom (1979) suggests, feathers may have first evolved for other reasons such as increased insulation, display features for ritual combat, mate attraction or as an aid to facilitate the catching of prey (Figure 7.2). If the latter, feathers may also have helped their owner to leap up after small flying insects. It is not difficult to see that natural selection might favour increased size of feathers if they improved their owner's success in catching prey. With natural selection bringing about steady improvements in the leaping capacities of these predators,

Figure 7.2.   Steps in the evolution of flight. Feathers may have originally evolved from scales for reasons other than flight. One possible explanation is that they helped insulate the animal against heat loss. Another is that they may have aided ground-running predatory ancestors of birds to catch insects (above). Once these feathers were large enough, small leaps after prey could become larger leaps until short flights became possible (see text).

leaps could lengthen in duration and height and, eventually, leaps would be of sufficient duration to qualify as free flight. As an alternative scenario, feathers may first have aided gliding out of trees. If longer more controlled glides were adaptive, selection would favour increase in wing and tail feather length. It would also favour any tendency to flap the feathered arms as a means for sustaining and controlling movements in the air. In such a way, it is entirely possible that every intermediate condition between non-flight and flight could have been selected for and in so doing brought ancestral birds to the brink of free flight. No design or anticipated end result would have been required.

There are many examples known of living organisms that have begun to use a structure for some other purpose than that for which it appears to be primarily suited. Penguins' use of vestigial wings and wing feathers as paddles under water quite clearly does not explain why feathers and wings evolved in the first place. Another example would be the 'woodpecker' finch of the Galapagos Islands which uses its bill to hold a tool (a small twig or cactus spine) which it pokes into crevices in search of insects. The bill of this bird did not evolve in response to selection pressure to hold the twig.

Yet, if there were not other birds known to show that bills were mostly used for direct feeding rather than tool-holding, we might well wonder how bills could have evolved prior to their being substantial enough to wield a cactus spine. Clearly old structures can find new functions and thereby become subject to new natural selection pressures.

In addition to all of the evidence for the adaptive value of functionally intermediate structures is yet another well-known process that can lead to significant evolutionary change. This is the area of rate and/or genetically controlled chemical changes in embryological development. Gould (1977) reviews much of the evidence which demonstrates that slight changes in rate-controlling genes can have profound effects on the structure of organisms.

One of the best-known examples is the axolotl, a South American salamander. In its functionally adult condition it resembles the larval (unmetamorphosed) forms of other species of salamander. Unlike these other species, however, it matures sexually in what is in reality a larval state and in this form reproduces to produce more larval forms. Hence its body normally never metamorphoses into what previously would have been its adult form. This profound change in body form of a sexually reproducing organism appears to be the result of a simple change in embryonic development (Gould, 1977).

A less well-known case has long been advanced. Gould (1977) and Archer and Aplin (1984) review evidence for recognising in adult humans the features of juvenile great apes. These juvenile ape-like features include our flat faces, protruding chins (rather than forward-jutting and hence chin-less faces which occur in adult great apes), relatively lighter and less coarse body hair, ventrally oriented vaginas and so forth. These features which occur in juvenile chimpanzees are lost as they mature into adults. To explain why they may have been retained in humans, we only need to envisage a slight change in a rate-controlling gene (or genes) in our australopithecine ancestors. Such a relatively minor change could have resulted in a relative slowing down of our bodily development but not the time at which we reach sexual maturity.

This hypothesis would also explain our long lives relative to most other similar-sized mammals (we 'age' more slowly), long period of infant dependence because we are slow to mature, and our large brains because they continue to grow longer at the more rapid juvenile rate. Gould (1977) discusses most of the evidence for this highly plausible hypothesis. It would also explain why, despite striking biochemical similarity to the great apes (see Chapter 6), we appear in these features to be otherwise dissimilar to chimpanzees. In particular, it would explain why this dissimilarity is much greater in comparison with adult than juvenile chimpanzees.

Similarly, Thulborn (1985c) reviews experimental evidence which shows that chicken embryos can be induced to change what would have been scales into feathers and feathers into scales. One might expect that these embryological transmutations should result only from a major overhaul in the genetic instructions of the chicken. But this turns out not to be the case. In fact, scales and feathers are identical in chemistry, molecular structure and mode of development. They only differ in shape and it is now clear that determination of this shape may depend on some strikingly simple control factor. In fact, the laboratory transformation in embryos was triggered by a single chemical — retinoic acid, better known as vitamin A.

How improbable then is it that feathers evolved from reptilian scales by a simple change in the chemical environment of the embryo skin? The fossil record clearly indicates that reptiles evolved into birds and scales into feathers (see Chapter 6) so this embryological scale-to-feather transition should come as no surprise to anyone except perhaps Creation 'Scientists'.

The point is that although the fossil record does in fact abound with 'links' (Chapter 6), the apparent absence of particular intermediate forms may reflect relatively rapid morphological transitions induced by nevertheless simple mutations in rate-controlling genes or in genes that control the balance of the chemical compounds which nourish the developing embryo. This sort of transition has been called 'clandestine' evolution (Thulborn, 1985c) because it would occur without leaving the sort of evidence that could be preserved in the fossil record. It may be one of the evolutionary processes responsible for instances of what appear to be punctuated equilibrium in the fossil record (such as the evolutionary transition between species of elephantine elephants; Chapter 6).

The question of how natural selection could lead to complex biochemical processes via less complex intermediate systems is just as interesting. As long ago as 1945, N.H. Horowitz proposed a model to explain how complex metabolic systems may have evolved (Dickerson, 1978). Horowitz suggested that if today a series of metabolic steps goes from substance A to substance E via substances B, C and D, it is probable that the ancestral need was for substance E which existed as a free solute in the organism's environment. Evolution of the chemical cycle would have begun when substance E ceased to be readily available in the ancestral organism's environment perhaps because it was progressively used up from the residual 'prebiotic soup'. Selection would then have favoured the ability to utilise a pre-existing enzyme that could convert a similar but more readily available compound (e.g. substance D) into substance E. And on it would go, each step in the cycle being added as a means to the same end. In this way, the complex stages in a metabolic chain could develop and persist as a progressively more circuitous means to the same end.

For example, photosynthesis probably evolved in prokaryotic organisms that once depended on anaerobic fermentation of readily available high-energy compounds to obtain glucose. As the supply of these energy-rich compounds was steadily consumed, selection pressure would have become intense to find other means of obtaining glucose. This sort of biochemical evolution would have been opportunistic. The intracellular waste products (or excess) of one metabolic process could well have become the source material for the progressive evolution of others and in this way complex interrelated metabolic pathways would have evolved.

Of course this is speculative but there is nothing improbable, let alone impossible, about the idea. Yet, to Creation 'Scientists', complex metabolic pathways are incontrovertible evidence for the impossibility of evolution. That's a shame because the hypothesis for their natural origin is ever so much more fascinating (and demonstrable, depending as it does only on natural processes) than the hypothesis of supernatural Creation.

To summarise; the supposed problem of the improbability of intermediate conditions being adaptive is simply not a problem. For that matter, it is evident that in some cases (such as those involving mutations in rate-controlling genes) there is no need to presume that for evolution to occur intermediate structures had to have existed. The problem is Creation 'Scientists'' failure to recognise that there is nothing about evolutionary processes which determines that they must operate with a plan or direction, that they must proceed to some end along a straight line, or that evolutionary change had to proceed at a steady rate. Evolution is the result of basically opportunistic processes at work in a constantly changing world.

The age of the Earth

The age of this planet has been considered in Chapters 3 and 4. It is also highly relevant in discussions of evolution. Modern understanding of rates of evolutionary change require an age for the Earth far in excess of that given by Creation 'Scientists'. Belief in a literal interpretation of the Bible gives Creation 'Scientists' an age for the Earth (and universe) of about six thousand years (although some, such as Gish and Bliss (1981), allow up to twenty thousand). In contrast, nuclear physics, astrophysics, geophysics and geochemistry have led scientists to the conclusion that the universe is between 10 000 000 000 and 20 000 000 000 years old and the Earth at least 4 500 000 000 to 4 600 000 000 years old (Awbrey, 1983; Brush, 1982; Freske, 1980). The age of the Earth as calculated by geophysicists would be sufficient to enable the various processes of evolution to produce all of the life on Earth.

CREATION 'SCIENCE'S' CHALLENGES TO THE EVIDENCE OF THE FOSSIL RECORD FOR THE REALITY OF EVOLUTION

The fossil record is perceived by most Creation 'Scientists' as the biggest threat to acceptance of the Biblical account of Creation and perhaps because of this they spend most of their time attacking the evidence it provides for the model of evolution. That is a very big job because the known record, already enormous, continues to grow at a prodigious rate.

There are two essentially different challenges issued by Creation 'Scientists' to discredit the evidence of the known record. First, they make the general claim that nothing in the fossil record demonstrates the existence of links between known kinds of organisms. Second, they say that the disposition of fossils in space and time provides no support for the evolutionary interpretation that life evolved progressively but that it does provide support for the hypothesis that all of the basic kinds of life appeared on Earth simultaneously during one week sometime between 6000 and 20 000 years ago.

If Creation 'Scientists' are right, then the fossil record does not provide support for the evolutionary model and evolutionary biologists have been misled into thinking that it corroborates the clear evidence for evolution provided by study of living organisms. We will consider their challenges in order.

The problem of 'kinds' and 'links'

Are there or are there not links between kinds of organisms? Creation 'Scientists' say no; evolutionists say yes. Although the fossil record most definitely does provide evidence of intermediate forms (Chapter 6), Creation 'Scientists' struggle valiantly to convince themselves and others by sophistry that it does not. Their published reasons for discounting known transitional forms as links depend in large part on their redefinitions (or, more commonly, on their refusal to accept current understanding) of critical terms such as 'links' and 'kinds' of organisms. By keeping both of these concepts nebulous, Creation 'Scientists' have built around themselves an almost impregnable fortress of semantics.

The concept of 'kind' among Creation 'Scientists' has steadily evolved in a transparent effort to ensure that no evidence from the real world can threaten the literal word of Genesis (Awbrey, 1981; Cracraft, 1983; Ruse, 1982). Most Creation 'Scientists' will 'allow' post-Creation week development (they dislike the term evolution) *within* created 'kinds'. They know, by faith in the literal truth of the Genesis account, that none of the created 'kinds' could have arisen from any other because they were created as fully formed 'kinds'. To Creation 'Scientists', there cannot be links between the Biblical 'kinds' of organisms because the literal word of the Bible is infallible.

Originally Christians interpreted God's instruction in Genesis for creatures to bring forth progeny 'after their kind' as an indication that the Biblical 'kinds' were in effect biological species (i.e. those creatures that could successfully reproduce). Because there was nothing known from the fossil record to challenge this notion, Christians had no need to defend or protect it.

Later, after biologists and palaeontologists discovered a world full of evolutionary transformations, modern Creation 'Scientists' (Siegler, 1978) decided that the created 'kinds' had better be redefined as those creatures that could produce offspring (fertile or otherwise) amongst themselves thereby dismissing as 'variation within kinds' what were otherwise potentially threatening demonstrations of links between the Biblical 'kinds'.

Gish (1979: 32), evidently not satisfied with the armour of previous definitions, tried to define 'kind' in a different way; one that would provide a more impregnable semantic fortress. He suggested that: 'A basic animal or plant kind would include all animals or plants which were truly derived from a single stock.' By way of elaboration, he said that in some cases a 'kind' might be a species (e.g. *Homo sapiens*, naturally) while in others it might be a genus (e.g. all of the species of *Canis* including coyotes, foxes, dogs and wolves).

As it stands, Gish's definition is utterly vacuous because it could apply either to the whole of life or to a single family unit within a species. He does not provide any guide for determining 'kinds': 'We cannot always be sure, however, what constitutes a separate kind' (Gish, 1979: 34–35). In fact, in this sense, 'kinds' (e.g. Mammalia) can even *contain* other 'kinds' (the 'dog kind'). Again this is a thinly disguised attempt to put the underlying concept of the created integrity of basic 'kinds' beyond examination.

What Creation 'Scientists' are attempting to achieve here is a definition so broad that any demonstration from the fossil record of structurally intermediate forms as evidence for evolution can be dismissed. This is because they can declare the intermediate to be either an indication that the two otherwise dissimilar linked forms are in fact one previously unrecognised created 'kind' or a member of one of the two previously recognised 'kinds' which it appears to link. Either way, the newly discovered intermediate comes to serve merely as a demonstration of 'variation within a single kind'.

An example of this is the way Creation 'Scientists' treat the rich fossil record of hominines. This is a particularly critical example because all Creation 'Scientists' believe that God created human beings. So what do they do with the transitional sequences in time (see Figure 6.10) from *Australopithecus* to *Homo sapiens*?

Gish simply declares the species of *Australopithecus* and *Homo habilis* and *H. erectus* to be apes (Gish, 1979: 104–146). The only significant contribution Gish permits the fossil record to make to the early history of *Homo sapiens* is neanderthal man which no modern biologist or anthropologist considers represents anything other than an isolated early population of *H. sapiens*. To the committed Creation 'Scientist', the rich fossil record of manlike primates can be no threat to Biblical Creation because, as biochemist Gish points out to anthropologists, it contains just peculiar apes and modern man; there are no transitional forms linking modern humans to any other kind of creature. Obviously, no matter how clear the fossil record is for evolutionary transitions, Creation 'Scientists' cannot accept it because they must preserve the infallibility of the Book of Genesis as they understand it.

An interesting demonstration here of the capricious manner in which Creation 'Scientists' deal with embarrassing aspects of the fossil record is the contrast in conclusions reached by two notable Creation 'Scientists', Duane Gish (who in 1974 was Associate Director of the Institute for Creation Research) and Henry Morris (an hydraulic engineer who in 1974 was the Director of the same institute), about the significance of *Homo erectus* as an obvious link between modern man and older hominines. Gish's treatment, noted above and given in Gish (1974a), involved declaring this seemingly transitional form to be some kind of giant ape and nothing whatsoever to do with modern man. When faced with the same problem, Morris (1974a: 174) reached exactly the opposite conclusion: '*Homo erectus* was a true man, but somewhat degenerate in size and culture'. This degeneracy was attributed to '. . . inbreeding, poor diet and a hostile environment'. Although both Gish and Morris started with contemplation of a species that shared ape-like and human-like features, neither could accept the reality that stared both in the face. To make it go away, one of them concluded that *H. erectus* was undoubtedly merely a degenerate man while the other concluded that it was undoubtedly a peculiar ape. The possibility that it might represent an intermediate form between modern man and older hominines was clearly forbidden territory to both. The amusing fact that they leaped away from this reality in opposite directions merely serves to show that wishing the problem away was much more important than the explanation used by each as 'scientific' rationale for the leap.

As long as Creationists refuse to be specific as to what are (or to give any guidelines for determining what are) different 'kinds' of organisms, a 'kind' can be anything a Creationist says it is. This is how they can exorcise embarrassments such as *Homo erectus*. They make it impossible to demonstrate evolution because, by playing fast and loose with the facts, any such demonstration is promptly declared an example of variation within a created 'kind'.

Fortunately, some Creation 'Scientists', including Gish, have given examples of what are to them different created 'kinds'. For example, Gish (1979: 32–36) gives 38 examples of basic 'kinds' including the following: *Homo sapiens*; the dog 'kind'; collectively all of the species of finches in the genera *Geospiza, Camarhynchus* and *Cactospiza*; corn; protozoa; sponges; jellyfish; worms; snails; lobsters; bees; fishes; amphibians; reptiles; birds; herons; mammals; crocodiles; dinosaurs; pterosaurs; platypuses; bats; hedgehogs; monkeys; apes; gibbons; orangutans; chimpanzees; and gorillas. Even this list, however, demonstrates ambiguity in Gish's concept of 'kind'. Why, for example, are orangutans listed as a 'kind' separate from the ape 'kind', and both listed as 'kinds' separate from the mammal 'kind'? Ambiguities aside, however, the list at least provides examples of the 'kinds' Creation 'Scientists' presume were created by God.

On the surface, this willingness to itemise basic 'kinds' would seem to be

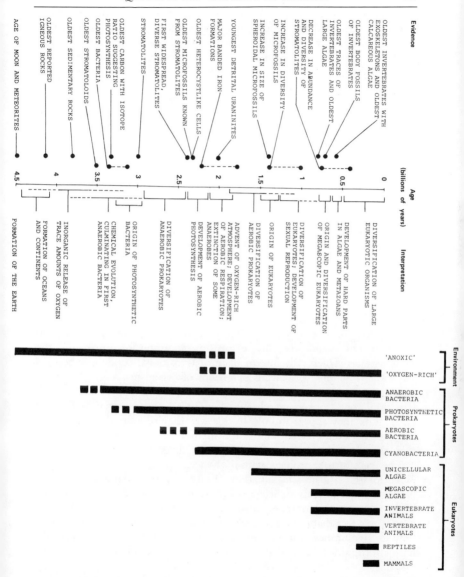

Figure 7.3 A summary of highlights of Earth history showing the ordered manner in which less complex forms of life (e.g. bacteria) appeared before more complex forms (e.g. animals). The evolution of cyanobacteria would have been correlated with production of sufficient oxygen eventually to alter the composition of the atmosphere from anoxic to oxygen rich. (modified after Schopf, 1978). One billion years is understood here to represent 1 000 000 000 years.

a risky move on Gish's part because he appears to have exposed a bit of Creation 'Science's' soft underbelly to the possibility of examination. With this list in mind, we should be able to consider whether or not there is evidence for believing that any of these are connected by transitional forms found in the fossil record to other 'kinds' of organisms. However, while it might appear that Gish has exposed Creation 'Science' to examination, using yet another semantic barricade, Creation 'Scientists' have ensured that it is still impossible to test their concept. This semantic obstacle is their refusal to define 'link' or 'transitional form' (Cracraft, 1983).

Failure to indicate precisely what it is that constitutes a 'link' between 'kinds' of organisms enables Creation 'Scientists' to declare that there are no such things. Gish (1979: 49) even says: 'As a matter of fact, the discovery of only five or six of the transitional forms scattered through time would be sufficient to document evolution.' Despite Gish's own awareness of transitional forms between reptiles and birds and between reptiles and mammals (Gish, 1979, 1980, 1981, 1985), he can make this statement because he knows that without defining what would constitute a 'link' he will not have to recognise any of the many intermediate fossil forms as 'links'.

What sort of organism or group of organisms should qualify as transitional forms or links? The *Macquarie Dictionary* defines link as: '1. one of the rings or separate pieces of which a chain is composed...anything serving to connect one part or thing with another...'. In an evolutionary sense, a link would be any organism or group of organisms that demonstrates structurally intermediate or transitional conditions between two other dissimilar organisms or groups of organisms.

Creation 'Scientists' dismiss all organisms that demonstrate intermediate structural conditions as merely unrecognised variation within one of the two 'kinds' they appear to link. In this way they reassure themselves that the gaps between created 'kinds' are impossible, by definition, to fill and that, therefore, there are no such things as 'links'.

There is a further problem with the concept of 'links' which often leaves evolutionists in a no-win situation. 'Links', by their nature, fit in gaps between 'kinds'. Creation 'Scientists' make as much of these gaps as they do of the supposed lack of transitional forms which define them. As a result, it is an ironic fact that each time a transitional form is discovered, no matter how its significance is dismissed by Creation 'Scientists', it automatically means recognition of two albeit smaller gaps where previously there was just one!

If we cut through the semantic armour used by Creation 'Scientists' to protect their beliefs, the fossil record presents unmistakable evidence for evolutionary transition between organisms at almost all taxonomic levels (see Chapter 6). From intermediates between the species in our own subfamily to intermediates between the classes of vertebrates, such as between reptiles and birds and between reptiles and mammals, the fossil record provides excellent support for the concepts and predictions of evolution. Conversely, it provides no support for the Creation 'Scientists'' hypothesis that all 'kinds' of organisms were made out of nothing 6000 to 20 000 years ago because the record does contain transitional forms between kinds of organisms which range from a few thousand to many millions of years in age.

The disposition of fossils in time: which model does it support?

The Creation 'Science' model for the history of life on Earth involves: (1) a beginning, some 6000 to 20 000 years ago (the age being calculated by adding up the lifespans involved in Biblical genealogies); (2) an initial period of six 24-hour days during which a supernatural being created, out of nothing, the whole of the universe including all living things; and (3) a worldwide Flood (the Noachian Deluge) which covered the whole of the Earth with water about 4000 years ago and, in so doing, produced within one year the Earth's sedimentary rocks and all of the fossil record 4000 years or older in age (Gish, 1979; Morris, 1974a).

In contrast, the evolutionary model involves: (1) the origin of life from non-life approximately 3 400 000 000 years ago; (2) the subsequent evolution of life forms from pre-existing life forms with a consequent increase in biological complexity through time (Figure 7.3); and (3) constant extinction of species since life's origin and steady but erratic accumulation of fossils since life first formed.

In Chapter 6, some of the abundant evidence from the fossil record in support of the evolutionary model has been reviewed. In contrast, scientists have discovered nothing in this record to support the Creation model.

Generally, Creation 'Scientists'' discussions of the fossil record focus on what they regard as the mistaken interpretations of evolutionists rather than on possible evidence in support of the Creation 'Science' model. The only notable exception to this is their claim for evidence in support of the co-existence of humans and a variety of organisms that palaeontologists regard as having become extinct long before humans evolved. Since the Creation 'Science' model involves sequential creation of all kinds of organisms in six 24-hour days, Creation 'Scientists' predict that the fossil record should reveal evidence for this co-existence. Humans should be found buried with dinosaurs and all other kinds of organisms because, not only did they live together, they all perished together in the Biblical Flood approximately 1600 years after the week in which they were all created.

The Paluxy River 'mantracks'
Creation 'Scientists' have claimed that at Glen Rose, Texas, in the limestone of the Paluxy River's bed, 'mantracks' co-exist in the same strata with

footprints of dinosaurs (Baugh, 1983; Beierle, 1977; Dougherty, 1971; Morris, 1980; Whitcomb and Morris, 1961a). The limestone is Early Cretaceous in age (about 105 million years old; although of course the actual age is disputed by Creation 'Scientists') and the dinosaur footprints appear to represent track types appropriate for this age. Several types of dinosaurs may have made these tracks (Cole *et al.*, 1985).

Because the Paluxy River 'mantracks' are cited by most Creation 'Scientists' as a demonstration of the support which the fossil record gives to their model, we should consider the evidence with some care. The photographs reproduced here (Figure 7.4) were taken by Creationists and said by one of them to offer the best evidence of giant mantracks in this area. What they *do* demonstrate, however, is how Creationists produced this best evidence. In (i) a legitimate dinosaur footprint (c), which is part of a dinosaur trackway running in the direction of the string (t), occurs adjacent to an almost two-foot-long 'mantrack' (a). Note that in the photograph the tracks were 'wet out with vaseline and alcohol for photographic purposes'. The 'mantrack', as 'wet out', shows a human-like appearance. If one were shown just this photo with its 'wet-out' 'mantrack' it would be difficult to fault the human-like appearance of the image. But note also the adjacent depression (b) which was not 'wet out' and the approaching water coming in from upstream (d). In (ii) the same 'wet out' 'mantrack' (a) is shown from slightly further away. The dinosaur trackway (e) is clear and the same individual print (c) is visible. Note also the sandbag (f) which has been placed on the adjacent rock in a futile effort to stop the water. In (iii), the fraud or wishful thinking of Creation 'Scientists' is clear. The water overran the 'mantrack' and it is now obvious that it was only a selectively 'wet out' part of a larger depression (a + b) which in fact is either a poorly preserved dinosaur track or a small scour mark at the end of a larger scour mark which led the water to the 'mantrack'. The human-like toes and sole painted distinctly in the first 'wet out' photograph appear to have vanished in the absence of the vaseline.

There have been many analyses by palaeontologists of the supposed 'mantracks' in this limestone (Bird (in Godfrey, 1981); Cole, 1985; Cole *et al.*, 1985; Cole and Godfrey, 1985; Godfrey, 1985; Hastings, 1985; Neufeld, 1975). The conclusions of these analyses, from the first by Bird in the 1930s to the most recent by Cole in 1985, can be summed up in the title of another analysis of Creation 'Scientists'' photographs: 'Dinosaur tracks, erosion marks and midnight chisel work (but no human footprints) in the Cretaceous limestone of the Paluxy River Bed, Texas' (Milne and Schafersman, 1983).

Movies made by Creation 'Scientists' including 'Footprints in Stone' have been similarly examined (Godfrey, 1981; Hastings, 1985; Milne and Schafersman, 1983; Weber, 1981). When scenes showing the supposed 'mantracks' are stopped for close inspection, they are found to be thoroughly unconvincing partly because it is impossible to distinguish the

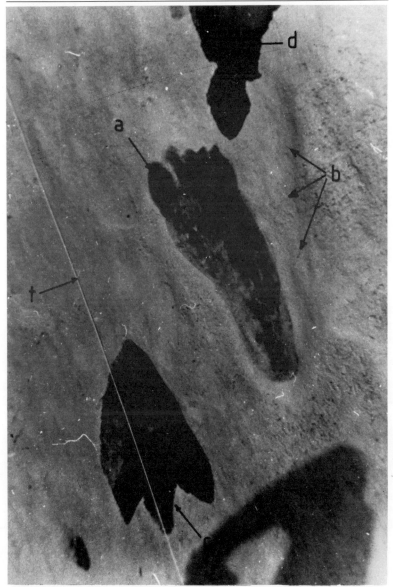

Figure 7.4(i). An example of Creation 'Scientists' idea of science — the Paluxy River dinosaur and 'mantracks'. (See text for detailed explanation.)

Figure 7.4(ii).

Figure 7.4(iii).

real structure of the mark from the superimposed layer of vaseline or shellac painted on the rock to 'reveal' or 'clarify' the outline.

In addition to these distortions many of the 'mantracks' show unmistakable signs of having been carved. Others are fraudulent misrepresentations of dinosaur tracks, scour marks, invertebrate burrow fillings and erosional features (Figure 7.4). Some of the supposed 'mantracks' exhibit insteps on the wrong side of the foot and claw marks jutting out from the 'heel'.

Not a single 'mantrack' has withstood serious examination.

Even some Creation 'Scientists' now admit the evidence is unconvincing (Milne and Schafersman, 1983; Morris, 1980: 146–147). Alfred West, who for several years helped Creation 'Scientist' Reverend Carl Baugh with his excavations of the river bed, came away convinced not only that there was no 'evidence' for human tracks but that Baugh knew this as well but was simply refusing to admit the fact (Hastings, 1985: 14). West also claims to have seen Creationist co-workers faking casts of the tracks to make them look more human-like.

Morris (1986: pp. ii–iii) has even stated in the Creation 'Science' publication *Impact* (No. 151):

> 'Due to an unknown cause, certain of the prints once labelled [by himself and other Creation 'Scientists'] human are taking on a completely different character. The prints in the trail which I have called the 'Taylor Trail', consisting of numerous readily visible elongated impressions in a left-right sequence, have changed into what appear to be tridactyl (three-toed) prints, evidently of some unidentified dinosaur. . . . In view of these developments, none of the four trails at the Taylor site can today be regarded as unquestionably of human origin. The Taylor Trail appears, obviously, dinosaurian, as do two prints thought to be in the Turnage Trail. The Giant Trail has what appears to be dinosaur prints leading toward it, and some of the Ryals tracks seem to be developing claw features, also. . . . The various controversial prints labelled as human by Carl Baugh in recent years are of uncertain origin. . .'

Morris concludes (1986: p. iv): 'Even though it would now be improper for creationists to continue to use the Paluxy data as evidence against evolution, in the light of these questions [about why the tracks are 'changing'], there is still much that is not known about the tracks and continued research is in order. We stand committed to truth, and will gladly modify or abandon our previous interpretation of the Paluxy data as the facts dictate.'

There have been other similar claims by Creation 'Scientists' that there are fossils 'out of place in time' (Kofahl, 1977; Morris, 1974a). All of these have been demonstrated to be either hoaxes and/or possibly deliberate misrepresentations by Creation 'Scientists' of legitimate finds (Conrad, 1981, 1982). For example, Reverend Boswell claimed in 1973, in an evolution/Creation debate, that the 'Meister Track' represented a boot print

TABLE 7.1
Summary of the key events in Earth's history

Time (MYA)*	Era	Period	Epoch	Events
0 to 65	Cenozoic	Quaternary	Pleistocene	Evolution of man
		Tertiary	Pliocene	
			Miocene	Extensive mammal
			Oligocene	radiations
			Eocene	
			Paleocene	First primates
65 to 144	Mesozoic	Cretaceous		Last dinosaurs First marsupials First flowering plants First monotremes
144 to 200		Jurassic		First birds Dinosaurs diversify
200 to 250		Triassic		First mammals and dinosaurs Therapsids dominant
250 to 286	Paleozoic	Permian		Major marine extinctions Pelycosaurs dominant First mammal-like reptiles
286 to 360		Carboniferous		First reptiles Arborescent lycopods Pteridosperms
360 to 408		Devonian		First amphibians Jawed fishes diversify
408 to 438		Silurian		First land plants
438 to 505		Ordovician		Burst of · diversification in metazoan families First fish
505 to 590		Cambrian		Oldest known bone First chordates Diverse metazoans

TABLE 7.1   Continued

590 to 700	Pre-Cambrian	Ediacaran	First skeletal elements Diverse soft-bodied metazoans First soft-bodied metazoans First animal traces (coelomates)

*MYA = million years ago

on top of a trilobite in 500 million year old shale. He said that this assessment had been supported by 'three laboratories around the world'. When the Utah Museum of Natural History, one of the institutions he claimed had supported the assessment, was approached for confirmation, the response (from Professor W. Stokes of the Department of Geological Sciences) was that although the trilobite was genuine the 'footprint' was nothing more than a depression caused by a spalled off piece of shale (Conrad, 1981).

The other four most often cited examples of 'humans out of place in time' are the Calaveras, Castenedolo and Olmo skulls and the so-called 'Guadeloupe man' (Howgate and Lewis, 1984; Kofahl, 1977). Creation 'Scientists' claim these examples of young humans in old rocks have been deliberately hidden by scientists because they do not support the conventional scientific understanding of when humans appeared on Earth. All four have been examined and, as Creation 'Scientists' well know but have chosen to ignore, found to represent entirely unsurprising finds. The Castenedolo bones were shown as long ago as 1957 to be markedly different from fossil bones obtained in the same deposit and were determined to be recent burials dug into Pliocene sands and clays (Boule and Vallois, 1957).

The significance of the Olmo skull has been completely misinterpreted by Creation 'Scientists'. It is in fact just what they say it is: a human skull cap in upper Pleistocene gravels. But it causes absolutely no embarrassment because it is a fossil specimen exhibiting a morphology precisely appropriate for its position in time (Conrad, 1982).

The Calaveras skull, in contrast, was a deliberate hoax (Conrad, 1982). This fact was determined as early as 1879 by fluorine analysis and confirmed by later investigations. Yet Creation 'Scientists' ignore this venerable debunking and claim that evolutionists are embarrassed by this skull (Kofahl, 1977; Morris, 1974a).

Guadeloupe Man, a modern-looking skeleton from supposed Miocene sediments on the West Indian island of the same name, was recently put forward by British Creation 'Scientists' (Cooper, 1983) as an example of establishment scientists hiding evidence contrary to the evolutionary model (Howgate and Lewis, 1984). In fact, the specimen was on public display in the British Museum for many years between 1882 and 1967. Besides, the sediments from which it was obtained were long ago determined to be relatively recent in age. Even the Creationist geologist Reverend William Buckland concluded in 1836 'There is no reason to consider these bones to be of high antiquity, as the rock in which they occur is of very recent formation, and so composed of agglutinated fragments of shells and corals which inhabit the adjacent water.' When pressed during question time at a defence of this so-called evidence in support of the Creation model, Cooper admitted that no one except he regarded the Guadeloupe deposit to be Miocene in age and of course, as a Creation 'Scientist', he didn't believe in the Miocene anyway (Howgate and Lewis, 1984).

The continued misuse of these examples by Creation 'Scientists' is a clear demonstration that these fundamentalists are a desperate lot. They need evidence from the fossil record to support their model but, failing to find it among the many thousands of legitimate fragments of Earth's history, they are quite willing gently to distort or bend the facts of others to suit the religion of theirs. It is an understandable, but decidedly dishonest, procedure.

Flood geology and the nature of the fossil record

According to the account in Genesis, Chapter 7, God caused the whole of the Earth to be covered with a flood in which 'the waters prevailed upon the earth an hundred and fifty days'. The time of this purge, calculated from the number of generations from Adam to Noah recorded in the Bible, was evidently approximately 1600 years after the week during which God made the universe. According to Creation 'Scientists', this Flood was also the source of the Earth's sedimentary rocks and the period during which the fossil record older than 4000 years accumulated (Gish, 1979; Morris, 1977; Whitcomb and Morris, 1961a).

It is clear that there is no scientific evidence for a worldwide flood at any stage in the Earth's history. It is also clear that Earth's sedimentary rocks formed over a vast period of time (Table 7.1). The oldest are approximately 3 700 000 000 years old while the youngest are still forming. These sediments include many formations not deposited by water (such as the massively thick deposits of desert sands) interspersed with some that represent marine deposits and others that represent freshwater deposits. Further, within these sequences are many successive erosional unconformities indicating major breaks (lasting millions of years) in the process of sediment accumulation.

In any case, the appearance and disappearance of the additional amount of water (4 400 000 000 km³) required to cover the Earth's mountains, which is over three times the amount (1 370 000 000 km³) presently contained in all of the Earth's oceans, would have imposed simply impossible constraints on the pre-Flood creatures of Earth and the inhabitants of the Ark during its journey (Soroka and Nelson, 1983). If that much extra water fell as rain, the pre-Flood Earth had to have had an atmospheric pressure about 840 times higher than it has now and an atmosphere which consisted of 99.9% water vapour (which would, incidentally, have been unbreathable). Further, from a thermodynamic point of view, because 2.26 million joules must be given up as heat for each kilogram of water condensed out of the atmosphere (Soroka and Nelson, 1983), that much water vapour condensing into rain would have raised the temperature of the Earth's atmosphere in excess of 3500°C during the time of the Flood. The consequences for the occupants of the Ark in what would have been a boiling ocean and unbreathable atmosphere bear thinking about. Even if the extra water welled up from within the Earth, the temperature of subsurface waters of this volume, because of their closer proximity to the hot mantle of the Earth, would have resulted again in oceans boiling away at temperatures of approximately 1600°C. Either way, Noah's geese would have been cooked.

Other absurdities inherent in the study of Biblical 'arkeology' (Moore, 1983; Schadewald, 1982) include the problem of survival of diseases. If all of the disease-causing organisms known to be obligate parasites of particular hosts survived the flood, Noah's lot had to be raging pesthouses. The humans aboard, for example, had to have all of mankind's specific diseases, fatal and otherwise, including (among dozens of others) measles, poliomyelitis, typhus, four different kinds of malaria, gonorrhoea, syphilis, smallpox, leprosy, typhoid fever and pneumococcal pneumonia. If they could have survived these diversely horrid diseases and infections, the Ark's besieged occupants would have developed immunities. The host-specific diseases would then have died out because of a lack of non-resistant hosts. Clearly, without resort to miracles (God or the Devil having created disease-causing organisms after the Ark grounded) which Creation 'Scientists' claim not to have to rely on to support the Genesis account, the whole thing becomes rather silly. If, on the other hand, they allow a second miraculous Creation to account for present-day diseases, there can no longer be even a pretence of science about the hypothesis because miracles are outside the realm of science. The third alternative explanation, that present-day disease-causing organisms evolved within the last 4000 years from benign ancestors which were on the Ark, demands acceptance by Creation 'Scientists' of more rapid and substantial evolutionary changes than even evolutionists would allow.

Biogeography, the study of the present and past distribution of organisms, presents yet another mountain of difficulties for the Creation 'Science' model. For this reason and quite understandably, it is carefully avoided as a topic in many hard-core treatises on Creation 'Science' such as those of Morris (1974a) and Gish (1979). On the other hand, it has provided powerful support for the evolutionary model (Moore, 1983).

If we listen to Creation 'Scientists' who do defend this aspect of Biblical literalism (e.g. Whitcomb and Morris, 1961a), we are told that God instructed Noah to build an Ark because, apart from the few individuals which would be saved in the Ark (the saved number of each 'kind' varying between two and seven, Genesis 6: 19–20 or 7: 8–9), he was going to destroy 'every living substance' (Genesis 7: 4). To this Ark, at least two of every kind of creature on Earth (from amoebas to aardvarks and gum trees to guavas) somehow found their way. Then, approximately a year after the Flood waters rose they subsided and every living thing then remaining on Earth hopped, crawled, flew, swam, burrowed, floated, rolled or whatever down the rugged mountain slopes of Mt Ararat in Turkey and from there dispersed to every habitable region on Earth. Let's consider what this post-Flood dispersal involved.

In the first place, recall that Creation 'Scientists' know that the vast bulk of the fossil record of every continent was deposited during the single year of the worldwide Flood. Curiously, however, every continent's fossil record contains older and younger fossils belonging to groups of organisms unique to that continent. For example, while representatives of some kinds of organisms like *Nothofagus* trees and lions occur in the fossil record of more than one (albeit adjacent) continent, Australia with New Guinea is the only continent whose 'Flood' sediments contain (as fossil representatives of the still-living families) kangaroos, koalas, thylacines, platypuses, echidnas, legless lizards, freshwater turtles, over sixteen families or subfamilies of birds and many other unique groups. In other words, 'post-Flood' Australia alone was repopulated by descendants of the organisms that prior to the Flood just happened also to be unique to Australia.

Similarly, South America (with central America) is the only continent whose fossil ('Flood') record contains, amongst many other distinctive groups still living only in that area, caenolestids (rat possums), microbiotheriids (fossil opossums), twelve families of caviomorph rodents (such as agoutis), cebid (New World monkeys) and callithricid primates (marmosets), five unique families of bats and many unique families of birds and reptiles. If this biogeographic pattern did not develop because the older forms evolved in South America into their living descendants, Creation 'Scientists' must be aware of some mysterious but non-miraculous process that would account for this strikingly non-random pattern in the same way it would need to account for the otherwise strikingly non-random Australian pattern.

These biogeographic patterns would be, in contrast, predictable outcomes of an evolutionary process. The evolutionary model would predict that organisms evolving through time would produce descendants that resembled their regional ancestors. No divine Flood, mysterious process or other miracle is required to explain this biogeographic data. Hence it comes as no surprise to evolutionists to discover that fossil koalas are only known

from Australia. Even Darwin realised that biogeography provided important support for what he called the 'Law of Descent'. As early as the 1830s,
he realised that the fossil animals of Australia most resembled the living
animals of the same continent. In the same way, he discovered that the
extinct animals of South America resembled the living animals of that
continent. This only seemed to make sense if the older animals were in fact
the ancestors of the younger animals. To Darwin and most modern
scientists, these data supported a model of evolutionary descent with
modification.

Turning to the Creation 'Science' model for an explanation of these data,
there are immediate problems in trying to envisage barriers within, on and
above the violent Flood waters that would result in such a strict non-
random sorting of fossils into the Flood sediments of each continent. In
particular, severe problems seem to arise when we try to visualise how this
sort of segregation could have controlled the movements of aquatic animals
(e.g. the platypuses which, despite their ability to swim, were mysteriously
confined to Australia) as well as flying organisms (e.g. many unique families
of birds and bats which, despite the ability to fly above the flood waters, are
confined to the fossil and modern record of particular continents). In
seeking non-evolutionary explanations for these biogeographic patterns,
we should also recall that Creation 'Scientists' declare that miracles are not
required, a disclaimer that sounds like a good start but which is in fact one
impossible to maintain when the Creation 'Science' model is considered in
any detail.

To consider just one example relevant to Australia, contemplate the lot
of the platypuses that left what must have been a carefully constructed
platypusary on the stranded Ark (for they could survive in nothing less
than this as many zoologists have demonstrated). In the first place, since
Creation 'Scientists' have declared that at the time Mount Ararat was an
active volcano (LaHaye and Morris, 1976), the platypus pair would have
had to survive intense heat and a poisonous atmosphere. Then, because
there are no platypuses (fossil or living) anywhere in the world except
Australia, the surviving pair must have made their way across the barren
post-Flood wastelands (remember, every living substance had been
destroyed) of Eurasia. Their journey would have been arduous. There was
no food, for the organisms comprising their diet would have had no time to
multiply and spread (platypuses require prodigious amounts of quite
specific types of food every day). Neither was there fresh water for drinking
or bathing (they constantly need to bathe in fresh water). This simple
requirement would have been frustrated since every body of water stranded
on the continents would have been filled with the foetid carcases of all kinds
of dead animals as well as a host of other pollutants.

Assuming that they survived the 24 000 km journey, on arrival in
Australia they faced starvation and dehydration in a land more thoroughly
devastated than Hiroshima. How did they survive to reproduce?

Plant biogeography presents no less daunting problems. Consider the cactuses of the Americas. Between Turkey and the Americas there are many deserts that would have been suitable for cactuses. But because there are no fossil or living cactuses in these intervening lands, all of the single-minded and diseased cactuses (recall that every kind of organism, including host-specific parasites, had its Ark representation) must have hitched a ride to South America without so much as setting a root to soil along the way. Perhaps they did this by clinging to the flanks of departing caviomorph rodents?

To make each implausible event sound possible, it becomes necessary to invent additional and progressively sillier stories for its support. Very soon, the whole scenario degenerates into an admission that miraculous interven-tion was in fact required (e.g. God made their feet burn-proof, took away their hunger, changed their feeding habits, gave them an unerring instinct to travel in one direction before and another direction after the Flood, etc.). It rapidly becomes obvious that without divine intervention, the Creation 'Science' model is unable to explain any of what would have otherwise been thousands of astounding odysseys.

Equally silly is the Creation 'Scientists'' explanation of why the fossil record demonstrates a progression within the geological column from prokaryotes, to simple eukaryotes, to invertebrates, to vertebrates; and within vertebrates from fish, to amphibians, to reptiles, to mammals (Figure 7.4). The reason proposed for this is that as the flood waters swept in, the more advanced forms of life were able to reach higher ground, were thus drowned last and therefore are found higher up in the fossil record. The other main factor responsible is said to have been 'hydraulic sorting' by the flood waters (Whitcomb and Morris, 1961a).

Both types of Creation 'Scientists'' explanation fall to shreds when examined. Why, for example, are extinct Cenozoic molluscs buried only above Cretaceous mammals? Could they run faster? Why are the enormous ceratopsian dinosaurs only buried above Triassic mammals but below Tertiary mammals? Was no human large or slow enough to get buried in the bottom sediments and no trilobite able to swim to the top of the water to be buried in the younger 'Flood' sediments? Why are toothed birds, including many highly efficient diving birds, only buried in Mesozoic rocks when a peculiar group of giant, flightless birds from Australia is only found in Cenozoic rocks? Why are the many thousands of streamlined Devonian fish buried only below Carboniferous and Permian trilobites? For the duration of the Flood were they all ordered by God to be clutched by bottom-dwelling crabs while the trilobites were allowed to cavort overhead to be buried in younger sediments? Why are flowering plants buried only above Triassic mammals? Were they all collected up by remorseful dinosaurs pursued by the Flood waters and carried until buried for the first time in the Cretaceous rocks? Was the same Flood that supposedly whirled away mountains and deposited all of the sediments visible in the Grand

Canyon so fond of Cretaceous dinosaur eggs that it buried many nests now found in Cretaceous rocks in Mongolia and North America without even moving an egg out of place? The many thousands of similar observations based on the known fossil sequence make it extremely difficult to take the whole Noachian Flood scenario seriously.

The hydraulic sorting explanation is equally ridiculous because, as even Morris (an hydraulic engineer) should know, an object's hydraulic drag is directly proportional to its cross-sectional diameter (Schadewald, 1982). This means that the smaller, simpler organisms should have been transported further and buried last while the dinosaurs (none before the Triassic), mammoths (none before the Pleistocene), whales (none before the Eocene) and all other larger creatures should have been buried first which is clearly not what the record reveals.

Gaps and the Timing of the 'Creation Week'

A final point should be made about the disposition of fossils in time. Although the fossil record provides increasingly good evidence for evolutionary transitions within and between many groups of organisms, about other transitions it has provided less information (Chapter 6). Many of the reasons for this are abundantly clear. In the first place, only a small percentage of fossil-bearing rocks have been adequately examined. This explains why 'links' are being continuously discovered (such as the skull of a fifteen million year old ancestral platypus which was discovered only as recently as 1985). However, we also see all around us as well as in the record of the rocks the evidence for erosion processes that can destroy vast quantities of older sediments with their fossil record, burial processes that can place older sediments out of reach, the existence of certain groups of organisms in environments unsuitable to fossilisation (such as the communities which live in highland forests or stony tablelands) and other groups such as soft-bodied creatures whose lack of hard parts make them less likely candidates for fossilisation even if sedimentary environments are nearby. As well, we have considered processes such as 'clandestine' evolution that can produce major changes in an organism's shape without involving the need for structurally intermediate forms. Thus the failure to find 'ancestral' forms for all kinds of organisms comes as no surprise to geologists and palaeontologists.

Creation 'Scientists' nevertheless point to the still-remaining 'missing links' or gaps in the record as proof positive that something is terribly amiss with the evolutionary model which is of course a logical *non sequitur*. As Creationist Gish himself pointed out in 1979, demonstration of just a few examples of 'links' between kinds in the fossil record would be quite sufficient to support the evolutionary model. As we have seen in Chapter 6, despite the protestations and sophistry of Creation 'Scientists', many such 'links' have indeed been discovered.

Ignoring this fact, Creation 'Scientists' propose a radically different interpretation of the fossil record. They are of course looking for evidence of a Creation 'week' and they clearly suggest that it is to be found in the rocks of the Cambrian Period which are conventionally dated at between 500 and 590 million years old. The relatively sudden appearance of many different kinds of organisms in the rocks of this age seems to them to be clear evidence for the literal 48 hours of Days 5 and 6 of Creation Week when God created out of nothing *all* of the kinds of animals and plants.

Unfortunately for the Creation 'Science' interpretation, there are a number of critically embarrassing problems. For example, although Creationists often do not tell the truth about this matter (Baker, 1983; Gish, 1974b; Kofahl, 1980), there are thousands of well-documented Precambrian fossils from rocks ranging in age from about 3 500 000 000 years old right up to the base of the Cambrian from many areas of the world (Figure 7.3 and Thulborn, 1985d). Similarly, although the oldest are single-celled organisms, many of the younger Precambrian organisms are simple coelenterates, annelids and primitive-looking arthropods. Some, such as the Late Precambrian *Rangea* from Africa, even appear to be links between other kinds of organisms (in this case between the Precambrian members of the Order Telestacea and the younger 'sea pens' of the Order Pennatulacea; Jenkins, 1985). Further, as more comes to be known about the appearance of the more complex life forms in the record, it is apparent that it was anything but sudden although the evolution of hard shells during the Early Cambrian in many groups certainly resulted in Cambrian organisms being the first to be commonly preserved.

In other words, the fossil record provides unmistakable refutation of the Creation 'Scientists'' hypothesis. Animals cannot have been created in the Cambrian if they were already in existence in the Precambrian. Despite the claims of the Creation 'Scientists' there are hundreds of much older Precambrian animals. The same would apply to *any* time period selected by these fundamentalists to represent Days 5 and 6 of the Creation Week unless they chose the time of the oldest. known organisms, the simple prokaryotes found in rocks about 3,500,000,000 years old (Figure 7.3).

But for the moment, let's allow the Creationists' claims to stand. If God did create all of the Earth's kinds of organisms at the beginning of the Cambrian (or whenever they wish it to be so), Genesis 1: 20–31, if taken literally, makes it clear that *all* kinds of organisms were created within two 24-hour days. And herein lies yet another dreadfully awkward problem for these fundamentalists. If all kinds of organisms were indeed created within the same two-day period, they should all have left traces in the oldest rocks which contain traces of the first organisms. If this time is taken to be the Cambrian, then in Cambrian rocks alongside the primitive crustaceans, trilobites, onychophorans, brachiopods, coelenterates and echinoderms we should find the bones of whales, ichthyosaurs, plesiosaurs, nothosaurs, sea snakes, turtles, diving birds, otters, seals, porpoises and dugongs just to name a few. Of course we do not. In fact, these forms do not appear until much later in the fossil record.

Before Creation 'Scientists' rest content with the idea that the as yet unfilled gaps in the evolutionary sequences of organisms are a fatal flaw in the evolutionary model, they should stop for a moment to consider what this sort of missing evidence would mean for the Creation model. Unlike evolutionary models, the Creation 'Science' model predicts that *every* kind of organism should have a fossil record as old as the oldest known organism. And we are not asking here for specific transitional forms; *any* individual of each kind would do. Without belabouring the point, it should be obvious that the gaps in the predicted record arising from the Creation 'Science' model vastly exceed those which still irk the evolutionary model.

Taking one example, Gish (1985) has recently declared that each of thirty-five orders of flying insects represent different created 'kinds'. Yet none of these has a fossil record which extends back beyond the Late Carboniferous. While evolutionists have predicted that a common ancestral form probably existed in the Late Devonian to Early Carboniferous, Creation 'Scientists' are stuck with the prediction that they were all in existence from Early Cambrian times. Totalling up the amount of pre-Late Carboniferous fossil evidence missing in terms of the two models (using Gish (1985) figure 2: pp. 62–63), we discover that if we let the missing record predicted by evolutionists represent 100 units, 1420 units are missing from the record predicted by the Creation 'Scientists'. If we extend this sort of argument to the record for all 'kinds' of mammals, the amount missing from the record predicted by the Creation 'Science' model would be many orders of magnitude higher than that which is missing from the record predicted by the evolutionary model. Clearly, 'gaps' in the record are an enormously greater problem for Creation 'Scientists' than they are for evolutionists.

If evidential support from the real world counts for anything in assessing the worth of competing hypotheses, and Creation 'Scientists' at least say that it should, the Creation 'Science' model for the origin of life on Earth would not be seriously entertained in preference to the evolutionary model by anyone who practises science.

SUMMARY

The most determined and (in spite of constant failure) enduring attempts to falsify the evolutionary model have come from the Creation 'Scientists'. Although they themselves do not regard the Creation model as testable, it can be, has been, and has failed.

The Creation model predicts, among other things, that the kinds of organisms were created by God as distinct types. The fossil record reveals, however, many extinct groups, structurally and temporally intermediate between kinds of organisms, constituting support for the evolutionary model but falsification for the Creation model.

The fossil record also reveals a non-synchronous distribution of major kinds of organisms up through the fossil record in a pattern that corroborates the evolutionary model but falsifies the Creation model.

Attempts by Creation 'Scientists' to demonstrate coexistence between humans and dinosaurs (for example) have failed because the evidence has been found to be faked or misinterpreted.

Creation 'Scientists'' attempts to demonstrate that evolution is impossible because it contravenes the second law of thermodynamics are fallacious because they ignore the fact that the second law only applies to systems that are closed. Life is an open system with energy continuously available in the form of sunlight.

Arguments by Creation 'Scientists' about the improbability of the natural origin of life (i.e. without the intervention of a supernatural Creator) have become extremely strained. There is rapidly accumulating experimental evidence to show that when primordial conditions of Earth are recreated in laboratories, monomers (e.g. amino acids), polymers (e.g. proteinoids comprised of over 200 amino acids) and even protocells (coacervate spheres and bacteria-like spheres made of organic molecules) form spontaneously. Some of the protocells even grow and reproduce. The only step that has not yet been spontaneously demonstrated is the evolution of DNA but all of the basic units of which this molecule is made have appeared spontaneously.

All other attempts of the Creation 'Scientist' to falsify the evolutionary model on logical grounds have similarly failed, including the argument that intermediate conditions could not be functional and hence transitions between kinds of organisms could not have occurred. The mammal-like reptiles, with their double jaw-joints (the reptilian articular–quadrate system functioning alongside the mammalian dentary–squamosal system) falsify this Creationist assertion.

Alternative hypotheses proposed by Creation 'Scientists' as explanations for the fossil record, such as the Biblical Flood model, involve patently ridiculous assumptions (e.g. the hydraulic sorting apologetic) to explain aspects of the actual record that contradict what would be logical predictions of the model. These assumptions and the predictions of the Biblical Flood model have been tested and falsified. The record provides abundant support for the evolutionary model but nothing but contradictions for the Biblical Flood model.

Historical and modern biogeography also provide powerful support for the evolutionary model but overwhelming problems for the Creation 'Science' model. The Flood scenario will not work unless miraculous interventions enabled all of the kinds of organisms (from platypuses to pineapples) and all of the world's complex ecosystems to be re-established in their original homelands within 4000 years of the Ark's landing on the flanks of a volcano in Turkey. Since miracles are not examinable using scientific methodology, this becomes yet another demonstration that the Creation 'Science' model is faith in the literal truth of the Bible; it is not science. To make it work, because it must with their assumptions, nonsense as well as non-science must be woven tightly throughout the Creation 'Science' model.

While it is true that nothing in science can ever be proven to be true (for by what standard could we recognise truth even if we had it?), competing hypotheses can be tested. When this happens and one is falsified, it is part of the scientific method to set the dud on the shelf and get on with testing the ones that still remain viable. That approach to science is precisely what the Creation 'Scientists' do not do. Every aspect of the Creation model that could be tested has been tested, and falsified. Despite this fact, Creation 'Scientists' cling to the whole of their hypothesis as if it had never been tested. That is why Creation 'Scientists' are not scientists and why what remains of Creation 'Science' (the untestable fragments) is not science. They are deeply religious fundamentalists, committed to the proposition of the literal truth of the Bible, tied to a belief system that will not permit them to recognise its total lack of substance in the real world.

Henry Morris, the Director of the Institute for Creation Research, sums up the Creation 'Scientists'' most conclusive evidence for rejecting the evolutionary model as follows (Morris, 1967: 55): 'The final and conclusive evidence against evolution is the fact that the Bible denies it. The Bible is the Word of God, absolutely inerrant and verbally inspired. . . . The Bible gives us the revelation we need, and it will be found that all the known facts of science or history can be very satisfactorily understood within this Biblical framework.'

The simple choice for those curious about the natural world is between this pseudoscientific strait-jacket and an open mind.

ACKNOWLEDGEMENTS

The greatest debt for the urge to produce my contributions to this volume I owe to Drs Duane Gish, Henry Morris, Gary Parker and the thousands of other Creation 'Scientists' who know that they are right and that there can exist no evidence that would threaten this absolute certainty. These Biblical fundamentalists have managed to purge their minds of centuries of real scientific achievement as well as Christian scholarship and, in so doing, to place their minds well beyond the reach of reason. This represents an impressive leap backwards in time to a period in human history when the universe was rarely understood outside the guidelines allowed by superstition. Their intentional conflation of the methods and aims of science with those of religion, their persistent campaigns aimed at inadequately aware educationalists to force their particular brand of Biblical fundamentalism into science classes, their attempt to blackmail theists by declaring belief in God and acceptance of evolution to be incompatible and their presumption to determine for God the extent of His powers and the extent of His Creation represent four powerful rational and Christian reasons for confronting this pseudoscientific movement.

That primary acknowledgement aside, I would like to thank the many students, teachers and colleagues who have read and provided constructive criticisms of earlier drafts of this chapter. In particular, Professors Ted Thompson, John Maynard-Smith, David Sandeman, Ross Crozier, David Oldroyd; Drs Pat Dixon, Alex Ritchie, Barry Richardson, Frank Burrows, Bob Selkirk, Ms Suzanne Hand and Mr Henk Godthelp provided many useful comments and corrections. Dr Crozier encouraged use of the concept about criteria that would have to be satisfied for evolution *not* to occur (Chapter 2). Ms Jenny Taylor produced many of the diagrams and photographs.

FURTHER READING

Tactics used by Creation 'Scientists' to falsify concepts of evolution

Bridgestock, M. (1985) Ten checks upon creation science. *Australian Science Teachers Journal* 30(4): 26–32.
Cavanaugh, M.A. (1985) Scientific creationism and rationality. *Nature* (Lond.) 315: 185–188.
Freske, S. (1981) Creationist misunderstanding, misrepresentation, and misuse of the Second Law of Thermodynamics. *Creation/Evolution* 4: 8–16.
Godfrey, L.R. (ed.) (1983) *Scientists confront creationism*. W.W. Norton, New York.

Specific attacks used by Creation 'Scientists' outlined

Gish, D.T. & Bliss, R.B. (1981) Summary of scientific evidence for creation. *ICR Impact Series* 96.
Godfrey, L. R. (ed.) (1983) *Scientists confront Creationism*. W.W. Norton, New York.

Refutations of attacks by Creation 'Scientists'

Anon. (1978) Evolution. *Scientific American* 239.
Ayala, F.J. & Valentine, J.W. (1979) *The theory and process of organic evolution*. The Benjamin/Cummings Publishing Co., Menlo Park.
Cherfas, J. (ed.) (1982) *Darwin up to date*. IPC Magazines Ltd (as 'A New Scientist Guide'), London.
Futuyma, D.J. (1983) *Science on trial — the case for evolution*. Pantheon Books, New York.
Gingerich, P.D. (1983) Evidence for evolution from the vertebrate fossil record. *Journal of Geological Education* 31: 140–144.
Godfrey, L.R. (ed.) (1983) *Scientists confront Creationism*. W.W. Norton, New York.
Kitcher, P. (1982) *Abusing science: the case against creationism*. Massachusetts Institute of Technology Press, Cambridge, Mass.
Mayr, E. (1982) *The growth of biological thought*. The Belknap Press of Harvard University Press, Cambridge, Mass. 974 pp.
Miller, K. (1982) Answers to the standard creationist arguments. *Creation/Evolution* 3: 1–13.
Montagu, A. (ed.) (1984) *Science and creationism*. Oxford University Press, Oxford. 415 pp.

Newell, N.D. (1982) *Creation and evolution: myth or reality?* Columbia University Press, New York.

Patterson, C. (1978) *Evolution.* British Museum (Natural History), London & the University of Queensland, Brisbane. 197 pp.

Smith, J.M. (ed.) (1982) *Evolution now: a century after Darwin. Nature* and Macmillan Press Ltd, London. 239 pp.

Stanley, S.M.G. (1979) *Macroevolution, pattern and process.* W.H. Freeman, San Francisco. 332 pp.

Hypothesis that all life was created in one week

Creation 'Science'

Gish, D.T. (1979) *Evolution the fossils say no!* Creation-Life Publishers, San Diego. 189 pp.

Gish, D.T. & Bliss, R.B. (1981) Summary of scientific evidence for creation. *ICR Impact Series* 96.

Scientific

Cavanaugh, M.A. (1985) Scientific creationism and rationality. *Nature* (Lond.) 315: 185–188.

Cherfas, J. (ed.) (1982) *Darwin up to date.* IPC Magazines Ltd (as 'A New Scientist Guide'), London.

Cole, J.R., Godfrey, L.R. & Schafersman, S.D. (1985) Mantracks? The fossils say *no! Creation/Evolution* 5: 37–45.

Futuyma, D.J. (1983) *Science on trial — the case for evolution.* Pantheon Books, New York.

Gingerich, P.D. (1983) Evidence for evolution from the vertebrate fossil record. *Journal of Geological Education* 31: 140–144.

Godfrey, L. R. (ed.) (1983) *Scientists confront creationism.* W.W. Norton, New York.

Godfrey, L.R. (1985) Foot notes of an anatomist. *Creation/Evolution* 5: 16–36.

Kitcher, P.G. (1982) *Abusing science: the case against creationism.* Massachusetts Institute of Technology Press, Cambridge, Mass.

Miller, K. (1982) Answers to the standard creationist arguments. *Creation/Evolution* 3: 1–13.

Montagu, A. (ed.) (1984) *Science and creationism.* Oxford University Press, Oxford. 415 pp.

Newell, N.D. (1982) *Creation and evolution: myth or reality?* Columbia University Press, New York.

Spontaneous development of life

Anon. (1980) *Man's place in evolution.* British Museum (Natural History), London.

Cairns-Smith, A.G. (1985) The first organisms. *Scientific American* 252(6): 74–82.

Doolittle, R.F. (1983) Probability and the origin of life. *In* Godfrey, L.R. (ed.) *Scientists confront creationism.* W.W. Norton and Co., New York: 85–97.

Field, R.J. (1985) Chemical organization in time and space. *American Scientist* 73: 142–150.

Fox, S.W. & Dose, K. (1972) *Molecular evolution and the origin of life.* W.H. Freeman, San Francisco. 359 pp.

Margulis, L. (1981) *Symbiosis in cell evolution.* W.H. Freeman, San Francisco.

Mayr, E. (1982) *The growth of biological thought.* The Belknap Press of Harvard University Press, Cambridge, Mass. 974 pp.

Patrusky, B. (1984) Before there was biology. *Mosaic* 15(6): 10–17.

Schopf, J.W. (1978) The evolution of the earliest cells. *Scientific American* 239(3): 84–102.

Scott, J. (1981) Natural selection in the primordial soup. *New Scientist* 89: 153–155.

Evidence for later evolution

Archer, M. & Clayton, G. (1984) *Vertebrate zoogeography and evolution in Australasia.* Hesperian Press, Perth. 1203 pp.

Badgley, C.E. (1984) Human evolution. *In* Broadhead, T.W. (ed.) *Mammals.* University of Tennessee Department of Geological Sciences Studies in Geology 8, Knoxville: 182–198.

Gingerich, P.D. (1976) Paleontology and phylogeny: patterns of evolution at the species level in Early Tertiary mammals. *American Journal of Science,* 276: 1–28.

Gingerich, P.D. (1983) Evidence for evolution from the vertebrate fossil record. *Journal of Geological Education* 31: 140–144.

Gingerich, P.D. (1984) Primate evolution. *In* Broadhead, T.W. (ed.) *Mammals.* University of Tennessee Department of Geological Sciences Studies in Geology 8, Knoxville: 167–181.

Gould, S.J. (1980a) *Ever since Darwin: reflections in natural history.* W.W. Norton, New York. 285 pp.

Gould, S.J. (1980b) *The panda's thumb: more reflections in natural history.* W.W. Norton, New York. 343 pp.

Gould, S.J. (1983) *Hen's teeth and horse's toes.* W.W. Norton, New York.

House, M.R. (ed.) (1979) *The origin of major invertebrate groups.* Academic Press for the Systematics Association, London. 515 pp.

Johanson, D.C. & Edey, M.A. (1981) *Lucy: The beginnings of humankind.* Simon & Schuster, New York.

Kemp, T.S. (1982a) The reptiles that became mammals. *In* Cherfas, J. (ed.) *Darwin up to date.* IPC Magazines Ltd, London: 31–34.

Kemp, T.S. (1982b) *Mammal-like reptiles and the origin of mammals.* Academic Press, London. 363 pp.

Kermack, D.M. & Kermack K.A. (1984) *The evolution of mammalian characters.* Croom Helm, London.

Patterson, C. (1978) *Evolution.* British Museum (Natural History), London & the University of Queensland, Brisbane. 197 pp.

Szalay, F.S. & Delson, E. (1979) *Evolutionary history of the primates.* Academic Press, New York. 580 pp.

Washburn, S.L. (1978) The evolution of man. *Scientific American* 239: 146–154.

Wood, B. (1976) *The evolution of early man.* Eurobook, London.

Appearance of mutations resulting in increased fitness

Chao, L. & Cox, E.C. (1983) Competition between high and low mutating strains of *Escherichia coli. Evolution* 37: 125–134.

Dykhuizen, D.E. & Hartl, D.L. (1983) Selection in chemostats. *Microbiological Reviews* 47: 150–168.

Mortlock, R.P. (1983) Experiments in evolution using microorganisms. *BioScience* 33: 308–313.

CHAPTER EIGHT

EVOLUTION AND CHRISTIAN BELIEF

Ian Falconer

In this chapter I would like to explore the view of Creation and evolution seen by myself as both Christian and scientist. To set the context of the divergent attitudes of Christian groups on this issue it is necessary to review the history of the Darwinian controversy in the last century. It is also important to explore the 'Fundamentalist' position on Biblical interpretation. I conclude with a brief synthesis of present scientific understanding of the origin of the world, life and evolutionary change with basic Christian belief.

THE HISTORICAL SCHISM — DARWINIAN EVOLUTION VERSUS THE CHURCH IN VICTORIAN ENGLAND

Darwin was a very cautious and careful scientist. He started his career as the young biologist of the *Beagle*, believing, along with his captain Robert Fitzroy, in 'the Immutability of Species'. His belief was based on the Biblical concept of a once-and-for-all creation of all kinds of plants and animals which remained basically unchanged. This was a reasonable understanding of speciation since our eyes tell us that living creatures reproduce similar offspring which have differences that are small compared with the differences between species. However, even in the early nineteenth century there was a small number of geologists and biologists who were coming to the view that some sort of evolutionary change over a very long period of time was the origin of the great variety of life on Earth.

By the time Darwin returned from his voyage his ideas had changed. He started his first notebook on the 'mutation of species' as soon as he arrived home (Mellersh, 1968) but it was twenty-three years before he published *On the Origin of Species*. Reluctantly, he was almost pushed into print by parallel ideas coming from another biologist, Alfred Russel Wallace. Wallace sent to Darwin his essay *On the tendency of varieties to depart indefinitely from the original type* which was so close to Darwin's thought that the two men presented a joint paper to the Linnean Society in London in 1858. The following year Darwin published *On the Origin of Species* and in 1871 *The Descent of Man*. The ideas put forward were met with considerable enthusiasm by the scientific community and rapidly spread internationally.

However, it opened a wide breach with orthodox views on the literal interpretation of the first chapters of Genesis in the Bible. It also raised the spectre of man as the product of a soul-less survival of the fittest and related to apes instead of a special Creation in the image of God.

Antagonism of conservative groups soon came to the surface. A public clash occurred at the Oxford meeting of the British Association for the Advancement of Science in 1860. Meetings of the British Association were, and still are, attended by interested lay people as well as professional scientists. At them, controversial topics are aired in public forum. On this occasion a number of eminent people were present; Admiral Fitzroy (captain of the *Beagle*), Samuel Wilberforce (the Bishop of Oxford), Sir Richard Owen (an anatomist who opposed Darwin's ideas), Thomas Huxley and Joseph Hooker who both vigorously supported Darwin. Darwin himself was not present. After a rather dull paper entitled 'The Intellectual Development of Europe with Reference to the Views of Mr Darwin and Others' proceedings were enlivened by restless students who shouted down the next two speakers. The third started to speak about man and monkey and was also shouted down. At this point Bishop Wilberforce rose and spoke eloquently for half an hour attacking Darwin's ideas and 'proof'. At the end he resorted to ridicule. Turning to Huxley he asked whether it was through his grandfather or his grandmother that he claimed descent from a monkey.

Huxley was initially offended but then saw his opportunity and said to Hooker sitting next to him 'The Lord hath delivered him into my hands'. He rose and spoke quietly, defending Darwin's theory. He concluded by saying that he would not be ashamed to have a monkey for an ancestor but he would be ashamed if it were a man 'who prostituted the gifts of culture and eloquence to the service of prejudice and falsehood'.

This, as might be expected, brought the discussion into temporary chaos. With such a start to the public controversy it is hardly surprising that the battle is still being fought. However, even in Victorian England, the leading natural scientists soon accepted the inherent logic of the theory and the clear geological and biological evidence which supported it. Darwin was honest about the implications of his theory for human descent which he expressed in *The Descent of Man*. He realised the difficulties of the absence of an understanding of the mechanism of heredity. It has taken almost a century and the discovery of the structure of DNA to expand knowledge of the process. The other problem was the incomplete nature of the fossil record. Despite a century of very successful palaeontology this is still used as a weak point by those who attack the 'proof' of evolution.

Darwin's ideas were continuously criticised throughout his life though supporting evidence continued to accumulate. By the time of his death in 1887, however, he was accepted as a great pioneer of science and was buried with full honour next to Sir Isaac Newton in Westminster Abbey.

THE FUNDAMENTALIST CHRISTIAN POSITION

Throughout the centuries of Christian scholarship there have been two divergent forces in operation. At one extreme we have a view which uses the basic material of the Gospels as an intellectual and spiritual framework to be interpreted in the light of the knowledge of the day and the experience and insight of the individual. At the other there are those who adhere to an interpretation of both Old and New Testaments on the most literal basis possible revising their views only insofar as translations and manuscript sources improve the quality of the source material and scholarship permits the interpretation of difficult passages.

Both of these Christian approaches lead to problems. The first, which can be called liberal theology, can move so far from the Biblical source material that it is barely recognisable as Christian. It merges with other general philosophies against which it has to stand or fall on its own merits. The second approach, which is presently called 'Fundamentalist' or 'Conservative Evangelical', has the inherent problem that it accepts as literal truth some Biblical material of poetic, mythological and spiritual origin and therefore not readily interpretable in any literal sense. The fundamentalist Christian groups are particularly strong in Old Testament theology with an emphasis on Genesis. As a consequence, the clash between Darwinian evolution and the literal interpretation of Biblical Creation becomes particularly acute. The obvious problem is the irreconcilability of the literal understanding of Biblical Creation with its discrete phases and short duration and the very protracted and mechanistic process of evolution in which God has no clear role.

Exacerbating this problem is the theological concept of sin being caused by mankind's (Adam's and Eve's) weak nature and disobedience to God (Genesis, Chapter 2). The theme recurs in Paul's letter to the Romans, Chapter 5. There he explains the Jewish understanding of sin as being brought into the world by one man, who thereby also brought death into the world. The action of this man (Adam) in disobeying God's direct command resulted in God's condemnation of all mankind from thenceforth. Paul then explains that Jesus Christ, through his sacrificial death on behalf of us all, reconciled mankind to God once again.

Thus any questioning of the literal interpretation of the Creation story in Genesis is regarded by the most extreme fundamentalist groups as an attack on both the power and majesty of God in Creation and the grace brought to sinful mankind through Christ. The great majority of evangelical Christians, however, are prepared to view Chapters 1 and 2 of Genesis as crucially important for the understanding of God's purpose but not as scientific descriptions of events.

THE SCHOLARLY CHRISTIAN POSITION

Mainstream scholarly Biblical studies incorporate into Biblical interpretation as much knowledge from outside sources as is available and relevant. In particular, archaeological research of the last three decades has uncovered much detail of the Bronze Age civilisation in the Eastern Mediterranean region. From this and other accounts, strong support for the overall historicity of the Old Testament account has been obtained which is acceptable to both scholarly and evangelical groups.

Many of the towns mentioned in Biblical histories have been identified and some have been the site of recent 'digs'. One example of considerable interest is the ancient city of Jericho which was invaded by the Jewish tribes under the leadership of Joshua. The Book of Joshua is part of a larger historical body of knowledge describing events around the thirteenth century BC. Jericho is the oldest inhabited city known to present archaeology. The earliest human habitation on this site dates from about 8000 BC. About a thousand years later the inhabitants built a massive wall, a stone guard tower and an enormous ditch in the rock surrounding the city. Around the sixteenth century BC the city was destroyed by fire. It was re-occupied from the fourteenth to thirteenth century BC, when Joshua destroyed it. The site then remained uninhabited until the seventh century BC. The Jewish story of this conquest was probably not recorded in the present Biblical version until the sixth or seventh century BC. However, reference is made to the Book of Jashar which originated no later than the conquest period.

The early chapters of Genesis dealing with the Creation story are considered by some liberal scholars to be derived from two tablets, the first dealing with the origins of the cosmos and the second with the origins of mankind. The transition between the accounts occurs at Genesis 2:4 (Graves and Patai, 1964). The two stories of Creation are not duplicates, or even closely related. At the time of writing it was common scribal practice to record independent accounts, which were quite different in content, in the same portion of scripture. There are many examples of this practice in the Old Testament. Older stories of the Creation are found in the *Enuma elish*, of Babylonian origin, which has limited overlap with Genesis, Chapter 1. This account dates to about 2000 BC and has been discovered intact on seven cuneiform tablets of 156 lines each. The key difference between the two stories is the emphasis on one God who is all-powerful in Genesis and on a series of rival and vengeful gods in *Enuma elish*. In both stories the sequence of Creation events is tied to the order of planetary gods of the Babylonian week. There are other Biblical references to Creation in the Psalms, Isaiah, Job, etc. which give a different picture yet again: one much closer to Canaanite ideas. A further important creation story of antiquity, also from about 2000 BC, is the Sumerian *Gilgamesh Epic* which also starts with creation of a male human and animals but no woman. The man later falls in love with a priestess sent by the god Gilgamesh.

Thus the present scholarly position on the text of Genesis, Chapters 1 and 2, is that they represent two different sources emanating from the Hebrew tradition of the Creation of the world and the origin of mankind and evil. They are related to other ancient sources originating in the region in which the Jewish patriarch Abraham lived. It is apparent that they were not written as a scientific account of the Creation events and could never have been regarded by the ancient Hebrew people as factual, in the same way as King Solomon was factual. They do, however, show an appreciation of the nature of an all-powerful God and an understanding of mankind's ability to do what it knows is wrong. Christ's references to Genesis are particularly from the latter historical part. The only quotation from Genesis, Chapter 2, in the Gospels is an allusion to the ancient nature of the Jewish people.

Present-day theological scholars of almost all Christian denominations accept evolution as part of the general knowledge of the living world. They do not see any incompatibility between factual information derived from science and faith in God and belief in Jesus Christ. It is only the extreme 'Creationist' group, a very small but vociferous minority on the fringe of the Christian faith, which persists in driving a wedge between Biblical understanding and present-day knowledge.

CHRISTIAN ATTITUDES TO SCIENTIFIC KNOWLEDGE AND BIBLICAL SCHOLARSHIP

There has always been an ambivalence in the attitudes of Christian Church leaders to science. This is not surprising since much of 'science' prior to the Renaissance was astrology and closer to the occult than to logic or observation. During the revival of learning in Europe the Church authorities supported scientific investigation and such notable men as the astronomers Copernicus (a Catholic monk), Kepler (Lutheran) and Galileo (Catholic) were backed by their religious leaders. The physicist Newton (Church of England) was also a theologian but was regarded as holding unorthodox views. The arrest of Galileo by the Inquisition was largely political. His disturbing ideas of the Earth and planets rotating round the sun were thought to be an incitement to reject the authority of the Church which had until then concluded that the sun rotated round the Earth. The irony of this is that the official Church view was derived from the pagans Aristotle and Ptolemy. The rejected view came from the monk Copernicus who, working with Church finance and support, had, 75 years before, shown the planets circled the sun.

The Darwinian controversy was fuelled from both sides. As ardently as some Christians attacked evolution the atheist scientists of the nineteenth century used evolutionary theory to ridicule the Bible. After all, the literal interpretation of Genesis, Chapters 1 and 2, had been part of the backbone of Protestant theology (in particular) since the Reformation; to turn this into an ancient myth casts doubt on the veracity of the whole Bible. Fortunately many Christian scholars, especially such notable figures as Teilhard de Chardin (Jesuit priest and Professor of Geology; 1959) and A.N. Whitehead (one of the most influential of twentieth-century philosophers; 1978) have developed a theology which fully incorporates the concept of evolution and does not reject scientific knowledge.

In addition Christian Biblical scholarship has moved away from literalist interpretation of the Old Testament to a study of the independent sources out of which the whole Bible was woven. These sources contain traditional historical accounts, often from independent origins, throwing different light on the same events. They include sections of poetry, of philosophy and of myth. In this context myth can be described as a deep truth expressed in story form as if there had been a real happening. The value of a myth is not affected by whether its content is factual or not. It expresses symbolic truth. The Creation stories are regarded by most contemporary Old Testament scholars as essentially myth — but myth with a very important theological content. The essential content is that God is the basic founder of the universe and the Earth; the initial stage of this foundation was one of chaos and under God's power this was converted to order. The consequence of God's power on this planet was life including the development of mankind. God saw his Creation as essentially good. Genesis, Chapter 2, additionally has the message that mankind can do evil and has to accept responsibility for its actions.

In my view as a Christian, a God who is the source of order in the universe, who established the nature of matter and the physical laws governing its behaviour, can act through the living as well as the inanimate. Therefore, if we regard God as 'the ground of our being' (Tillich, 1955) we cannot exclude evolution from the range of God's power.

It has been argued that the whole evolutionary process is mechanistic and simply follows the laws of thermodynamics and chance. However, there are two issues which permit this view to be challenged. The first, simply expressed, is that it is not possible to deny a place for God in God's universe. For example, the molecular processes of evolution are rapidly being understood. Indeed the time is not very far off when a minimum form of living organism will be able to be made by the processes of biochemistry. This chemical creation of life, and indeed none of the areas of advancing science, will either disprove or prove the existence of God. To those who adopt a wholly mechanistic philosophy, the generation of 'life in a test-tube' will strengthen their views. To those who see God as the underlying power of all things, new scientific knowledge is an explanation, however partial, of the wonderful way God has operated to generate us and the world.

The second issue is that of behaviour. The difference between a life-form and an inanimate object is that the living organism responds to its environment. Even the most elementary life, such as the slime moulds, responds to hunger, light, temperature and humidity. The slime moulds behave in a way that helps them survive in adverse circumstances. They select where they go, stay as separate amoebae or clump into a mobile slug. They can form spores on a stalk and be blown large distances by the wind. Thus life-forms do not just obey the laws of chance, they actively respond to the circumstances in a way that helps them to survive. Mutation provides extra options which become especially important for survival when the environment changes as it has many times over the long history of the planet. Thus behaviour is purposive, and is part of the underlying reason for the development and evolution of life (Birch and Cable, 1981).

In mankind, it would appear that evolution has done something quite remarkable. It has directed the formation of a living organism which uniquely has the capacity to explore its own existence, the way in which it came to exist and the reasons for its existence. Now, it could be argued that evolution of philosophical thought of this kind was just the result of a mechanistic survival process for a species. However, a more convincing argument to me as a Christian and a scientist is that it arose as a behavioural response to man's search for the underlying purpose of life itself — in much the same way as behavioural responses to environmental stimuli arose in other species. In the case of evolving man the stimulus for the response was the increasing complexity of his life leading to the search for self-knowledge and thence the desire to understand life's purpose. The result of the response was the development of the ideas of God and man's relationship with Him. In its primitive Christian form, as expressed in Genesis, man is seen as a mirror of God. This view has gradually expanded as man's experience of God has expanded, particularly as revealed through the teachings of Christ. Now, man is seen as co-heir with Christ to an existence which extends beyond his temporal existence but arises from it.

SHOULD CHRISTIANS LOOK FOR PARTICULAR 'ACTS OF GOD' IN CREATION?

Now that most Christians accept that science provides a description of the mechanism for the functioning of the universe, the role of God in performing 'supernatural' acts becomes more difficult to understand. In the area of evolution it might be thought that the actual origin of life itself was a supernatural 'Act of God'. However, the beginnings of life on Earth are gradually unfolding to scientific investigation. The key pieces of information are the oxygen-free early atmosphere of the Earth, which allowed organic molecules to exist without oxidation, and the nature of DNA with its inherent ability for highly accurate replication. It seems likely that the primitive Earth environment and the inherent characteristics of the molecules involved in living organisms would, given time, produce life without postulating any event outside natural science.

Another area in which the role of God in evolution is worth thought is that of the origin of man's mind and soul. Here the soul can be considered as the eternal aspect of mind; that part of individual personality which transcends bodily death. All religions have a concept of an eternal aspect of the individual. Hinduism and Buddhism consider that all animal life possesses a spirit which has continuity through generations and that this spirit or soul achieves gradual perfection. In Christianity man is regarded as separate from the remainder of the animal kingdom but even so as man's mental powers increased his spiritual nature must have similarly developed.

It would thus seem difficult to support single-point Acts of God for the origin of life or the source of the soul. Both can be more convincingly defended on the basis of God providing structure and purpose for the universe, the consequences over time and through evolution being man as an animal and as a religious being.

EVOLUTION AS A KEY TO GOD'S PURPOSE

Because of the Fundamentalist objections to the concept of the descent of man from non-human species, the conceptual importance of evolution in Christian theology has been neglected.

In theological terms the purpose of mankind is to know and serve God. Since we are a reflection of God and the world a reflection of his design it follows that to achieve a greater knowledge of Him we must not only heed the teachings of God's people but also expand knowledge of ourselves and our world. The evolution of the mind is of critical importance to gaining this understanding.

The real factors that separate humanity from the rest of life on this planet are intelligence, ability to think in the abstract and means of conveying ideas to succeeding generations. Present evidence indicates that these attributes developed in our species within the last 100 000 years, perhaps as a result of the adverse environment of the ice ages. There are no reasons why other species could not evolve similar capabilities, should they have value for their survival. Perhaps on this planet the dolphins are on the track to the same mental abilities. Perhaps on planets of other stars life-forms are ahead of us in intelligence. Whatever the evolutionary situation elsewhere, on this planet mankind is the first form to develop thought to the extent that we can use mathematical abstractions to define relationships, or use experimentation to uncover the physical mechanisms of the universe.

Man has the ability to define philosophical and religious concepts; an ability which appeared earlier than either science or mathematics. The earliest written evidence for religious ideas is about 4000–5000 years old but, from its detail, it contains ideas clearly much older and handed down in unwritten form. The oldest evidences of religious ideas are perhaps the

Neolithic cave paintings in Europe and burials of about 20 000–30 000 years ago. These early European burials contained tools and food. China and Egypt also had ancient traditions of burial with the material necessities of life. These early cultures reflect man's great unwillingness to accept death as a finality, an unwillingness which may in part be due to the continuing power of the mind as the body ages towards death.

Mankind has had to evolve considerably from the African *Homo habilis* (living about two million years ago, with a brain about half the size of modern man; Sagan, 1977) before it could develop creative and philosophical capabilities. Without these evolutionarily derived features the whole development of religion is impossible. It required the brain of modern man to develop the concept of a Creator and of Creation, of a purpose for existence and even an idea of death. Thus for God to be known, species of sufficient intelligence need to be generated by evolution.

Thus I see evolution as both part of God's mechanism operating in the universe and part of God's purpose for the universe. I regard life itself — on this planet or on others — as an inherent and fascinating part of the existence of matter. The evolution of a species with self-knowledge, creative intelligence and the free will to use its power for good or evil points to a role for God beyond physics or molecular biology, indeed a role for God in the existence of man.

REFERENCES

Allin, E.F. (1975) Evolution of the mammalian middle ear. *Journal of Morphology* 147: 403–438.
Archer, M. & Clayton, G. (eds) (1984) *Vertebrate zoogeography and evolution in Australasia*. Hesperian Press, Perth. 1203 pp.
Archer, M. (1984) Origins and early radiations of mammals *in* Archer, M. and Clayton, G. (eds). *ibid*: 477–515.
Archer, M. & Aplin, K. (1984) Humans among primates: stark naked in a crowd. *In* Archer, M. & Clayton, G. (eds). *ibid*: 949–993.
Awbrey, F.T. (1981) Defining 'Kinds' — do creationists apply a double standard? *Creation/Evolution* 2: 1–6.
Awbrey, F.T. (1983) Space dust, the Moon's surface, and the age of the Cosmos. *Creation/Evolution* 4: 21–29.
Ayala, F.J. (ed.) (1976) *Molecular evolution*. Sinauer Associates Inc. Sunderland, Mass. 277 pp.
Badgley, C.E. (1984) Human evolution. *In* Broadhead, T.W. (ed.) *Mammals*. University of Tennessee Department of Geological Sciences Studies in Geology 8, Knoxville: 182–198.
Baker, S. (1983) *Bone of contention: is evolution true?* Third impression. Creation Science Foundation, Sunnybank, Queensland.
Bakker, R.T. (1983) The deer flees, the wolf pursues: incongruencies in predator–prey coevolution. *In* Futuyma, D.J. & Slatkin, M. (eds) *Coevolution*. Sinauer Associates Inc., Sunderland, Mass.
Bakker, R.T. (1985) Evolution by revolution. *Science* 85 6: 72–80.
Barnes, T.G. (1973) *Origin and destiny of the Earth's magnetic field*. Creation-Life Publishers, San Diego. 64pp.
Baugh, C. (1983) *Latest human and dinosaur tracks*. (Audiotape). Bible-Science Association Tape of the Month, Minneapolis.
Beierle, F. (1977) *Man, dinosaurs, and history*. Perfect Printing Co., Prosser.
Betz, J.L., Brown, P.R., Smyth, M.J. & Clarke, P.H. (1974) Evolution in action. *Nature* (Lond.) 247: 261–264.
Birch, C. & Cable, J.B. (1981) *The liberation of life*. Cambridge University Press. 353 pp.
Boag, P.T. & Grant, P.R. (1981) Intense natural selection in a population of Darwin's finches (Geospizinae) in the Galapagos. *Science* 214: 82–85.
Boule, M. & Vallois, H. (1957) *Fossil men*. The Dryden Press, New York.
Brace, C.L. (1983) Humans in time and space. *In* Godfrey, L.R. (ed.) *Scientists confront creationism*. W.W. Norton., New York: 245–282.
Brice, W.R. (1982) Bishop Ussher, John Lightfoot and the Age of Creation. *Journal of Geological Education* 30(1): 18–24.
Brush, S.G. (1982) Finding the age of the Earth: by physics or by faith? *Journal of Geological Education* 30(1): 34–58.
Bush, G.L. (1969) Sympatric host race formation and speciation in frugivorous flies of the genus *Rhagoletis* (Diptera, Tephritidae). *Evolution* 23: 237–251.
Bush, G.L. (1975a) Modes of animal speciation. *Annual Review of Ecology and Systematics* 6: 339–364.
Bush, G.L. (1975b) Sympatric speciation in phytophagous parasitic insects. *In* Price, P.W. (ed.) *Evolutionary strategies of parasitic insects*. Plenum Press, London: 187–206.
Carson, H.L. (1976) Inference of the time of origin of some *Drosophila* species. *Nature* (Lond.) 259: 395–396.
Clark, I.F. & Cook, B.J. (eds) (1983) *Geological science: perspectives of the Earth*. Australian Academy of Science, Canberra. 651 pp.
Colbert, E.H. (1969) *Evolution of the vertebrates*. John Wiley & Sons, New York.
Cole, J.R. (1985) If I had a hammer. *Creation/Evolution* 5: 46–47.
Cole, J.R. & Godfrey, L.R. (eds) (1985) The Paluxy River Footprint mystery — solved. *Creation/Evolution* 5: 1–56.
Cole, J.R., Godfrey, L.R. & Schafersman, S.D. (1985) Mantracks? The fossils say *no*! *Creation/Evolution* 5: 37–45.
Conrad, E.C. (1981) Tripping over a trilobite: a study of the Meister tracks. *Creation/Evolution* 2: 30–33.
Conrad, E.C. (1982) Are there human fossils in the 'wrong place' for evolution? *Creation/Evolution* 3: 14–22.

Cooper, W.R. (1983) Human fossils from Noah's Flood. *Ex Nihilo* 6(2): 31–32.

Cracraft, J. (1983) Systematics, comparative biology, and the case against Creationism. *In* Godfrey, L.R. (ed.) *Scientists confront creationism.* W.W. Norton., New York: 163–191.

Crompton, A.W. (1985) Cranial structure and relationships of the Liassic mammal *Sinoconodon. Zoological Journal of the Linnean Society* 85: 99–119.

Crompton, A.W. & Jenkins, F.A. (1979) Origin of mammals. *In* Lillegraven, J.A., Kielan-Jaworowska, Z. 8 Clemens, W.A. (eds) *Mesozoic mammals: the first two thirds of mammalian history.* University of California Press, Berkley: 7–58.

Curtis, H. (1983) *Biology.* 4th edition. Worth Publishers, New York. 1159 pp.

Dalrymple, G.B. (1983) Can the Earth be dated from decay of its magnetic field? *Journal of Geological Education* 31(2): 124–133.

Dalrymple, G.B. (1983) Radiometric dating and the age of the Earth. *In* The Creationist Attack on Science — Symposium. 1982. *Proceedings of the Federation of American Societies for Experimental Biology* 42(13): 3022–3042.

Darwin, C. (1859) *On the Origin of Species by Means of Natural Selection or the Preservation of Favoured Races in the Struggle for Life.* Murray, London.

Delson, E. (ed.) (1985) *Ancestors: the hard evidence.* Alan R. Liss., New York. 366 pp.

Dickerson, R.E. (1978) Chemical evolution and the origin of life. *Scientific American* 239(3): 62–78.

Dobzhansky, T., Ayala, F.J., Stebbins, G.L. & Valentine, W.H. (1977) *Evolution.* W.H. Freeman, San Francisco. 572 pp.

Doolittle, R.F. (1983) Probability and the origin of life. *In* Godfrey, L.R. (ed.) *Scientists confront creationism.* W.W. Norton, New York.

Dougherty, C. (1971) *Valley of the giants.* Bennett Printing Co., Cleburn.

Edwards, F. (1983) Is it really fair to give Creationism equal time? *In* Godfrey, L.R. (ed.) *Scientists confront creationism.* W.W. Norton., New York: 301–316.

Emery, R.J. & Schultze, H.–P. (1984) Vertebrate paleontology reviewed. *Geotimes,* February 1984: 14–16.

Fackerell, E.D. (1984) The Age of the Astronomical Universe. *Ex Nihilo Technical Journal* 1: 87–94.

Fakhrai, H., van Roode, J.H.G. & Orgel, L.E. (1981) Synthesis of oligoguanylates on oligocytidylate templates. *Journal of Molecular Evolution* 17: 295–302.

Feduccia, A. (1981) *The age of birds.* Harvard University Press, Cambridge.

Ferre, F. (1983) Science, pseudo-science, and natural theology. *Journal of Geological Education* 31: 83–86.

Fox, S.W. (1981) Origins of the protein synthesis cycle. *International Journal of Quantum Chemistry, Quantum Biology Symposium* 8: 441–454.

Fox, S.W. (1984) Proteinoid experiments and evolutionary theory. *In* Ho, M.–W. & Saunders, P.T. (eds) *Beyond neo-Darwinism.* Academic Press, London: 15–59.

Fox, S.W. & Matsuno, K. (1983) Self-organization of the protocell was a forward process. *Journal of Theoretical Biology* 101: 321–323.

Freske, S. (1980) Evidence supporting a great age for the universe. *Creation/Evolution* 1: 34–39.

Freske, S. (1981) Creationist misunderstanding, misrepresentation, and misuse of the Second Law of Thermodynamics. *Creation/Evolution* 2: 8–16.

Gallant, R.A. (1984a) To hell with evolution. *In* Montagu, A. (ed.) *Science and creationism.* Oxford University Press, New York: 282–305.

Gallant, R.A. (1984b). *Ibid.,* p. 292.

Gibby, M. (1981) Polyploidy and its evolutionary significance. *In* Forey, P.L. (ed.) *The evolving biosphere.* British Museum (Natural History) & Cambridge University Press, Cambridge: 87–96.

Gingerich, P.D. (1976) Paleontology and phylogeny: patterns of evolution at the species level in Early Tertiary mammals. *American Journal of Science* 276: 1–28.

Gingerich, P.D. (1983) Evidence for evolution from the vertebrate fossil record. *Journal of Geological Education* 31: 140–144.

Gingerich, P.D. (1984) Primate evolution. *In* Broadhead, T.W. (ed.) *Mammals.* University of Tennessee Department of Geological Sciences Studies in Geology 8, Knoxville: 167–181.

Gingerich, P.D. & Simons, E.L. (1977) Systematics, phylogeny, and evolution of Early Eocene Adapidae (Mammalia, Primates) in North America. *University of Michigan, Museum of Paleontology. Contributions* 24: 245–279.

Gish, D. (1973) *Evolution the fossils say No!* Creation-Life Publishers, San Diego. 129 pp.

Gish, D. (1974a) *Evolution the Fossils Say No!* Zonderman, San Diego. 77 pp.

Gish, D.T. (1974b) *Have you been brainwashed?.* Life Messengers, Seattle.

Gish, D.T. (1979) *Evolution the fossils say no!* Creation-Life Publishers, San Diego. 189 pp.

Gish, D.T. (1980) The origin of mammals. *Impact* No. 87: i–viii. (Published by the Institute for Creation Research.)

Gish, D.T. (1981) The mammal-like reptiles. *Impact* No. 102: i–viii. (Published by the Institute for Creation Research).

Gish, D.T. (1985) *Evolution: the challenge of the fossil record.* Creation-Life Publishers, El Cajon, California. 278 pp. (This is really another edition of Gish, 1979.)

Gish, D.T. & Bliss, R.B. (1981) Summary of scientific evidence for creation. *ICR Impact Series* 96.

Glaessner, M.F. (1961) Pre-Cambrian animals. *Scientific American* 204: 72–78.

Glaessner, M.F. (1984) *The dawn of animal life. A biohistorical study.* Cambridge University Press, Cambridge. 244 pp.

Godfrey, L.R. (1981) An analysis of the creationist film, *Footprints in stone. Creation/Evolution* 2: 23–30.

Godfrey, L.R. (ed.) (1983) *Scientists confront creationism.* W.W. Norton & Co., New York.

Godfrey, L.R. (1985) Foot notes of an anatomist. *Creation/Evolution* 5: 16–36.

Gosse, P.H. (1857) *Omphalos — An Attempt to Untie the Geological Knot.* Van Voorst, London.

Gould, S.J. (1977) *Ontogeny and phylogeny.* Harvard University Press, Cambridge, Mass. 501 pp.

Graves, R. & Patai, R. (1964) *Hebrew myths: the Book of Genesis.* Cassell, London. 311 pp.

Groves, C.P. (1985) Did human beings evolve? *In* Bridgestock, M. & Smith, K. (eds) *'Creationism': An Australian Perspective.* Mark Plummer for the Australian Skeptics, Melbourne.

Hallam, A. (1983) *Great Geological Controversies.* Oxford University Press, Oxford. 200 pp.

Ham, K. (1980) Creation science methodology. *Ex Nihilo* 3(2): 21.

Ham, K. (1983) The relevance of creation. Casebook II. *Ex Nihilo* 6(2): 2.

Harland, W.B., Cox, A.V., Llewellyn, P.G., Pickton, C.A.G., Smith, A.G. & Walters, R. (1982) *A geologic time scale.* Cambridge University Press, Cambridge. 131 pp.

Hastings, R.J. (1985) Tracking those incredible Creationists. *Creation/Evolution* 5: 5–15.

Holmes, A. (1937) *The age of the Earth.* Nelson, London. 263 pp.

Holmes, A. (1960) A revised geological time scale. *Transactions of the Edinburgh Geological Society* 17: 183–216.

Howgate, M.E. & Lewis, A.J. (1984) The case of Miocene man. *New Scientist,* 29 March 1984: 44–45.

Hyers, C. (1983) Genesis knows nothing of scientific creationism. *Creation/Evolution* 4: 1–21.

Jenkins, R.J.F. (1985) The enigmatic Ediacaran (Late Precambrian) genus *Rangea* and related forms. *Paleobiology* 11: 336–355.

Johanson, D.C. & Edey, M.A. (1981) *Lucy: the Beginnings of Humankind.* Simon & Schuster, New York.

Jones, J.S. (1981) An uncensored page of fossil history. *Nature* (Lond.) 293: 427–428.

Keast, A. (1958) The genus *Psophodes* Vigors and Horsfield, and its significance in demonstrating a possible pathway for the origin of Eyrean species from Bassian ones. *The Emu* 58: 247–255.

Kemp, T.S. (1982) *Mammal-like reptiles and the origin of mammals.* Academic Press, London. 363 pp.

Kermack, D.M. & Kermack, K.A. (1984) *The evolution of mammalian characters.* Croom Helm, London.

Kessler, S. (1966) Selection for and against ethological isolation between *Drosophila pseudoobscura* and *Drosophila persimilis. Evolution* 20: 634–645.

Kingsley, C. (1928) *The water-babies.* Macmillan, London.

Kitcher, P. (1982) *Abusing science: the case against creationism.* Massachusetts Institute of Technology Press, Cambridge, Mass.

Kofahl, R.E. (1977) *The handy dandy evolution refuter.* Beta Books, San Diego. 160 pp.

Kofahl, R.E. (1980) *Handy dandy evolution refuter.* Revised edition. Beta Books, San Diego.

LaHaye, T.F. & Morris, J.D. (1976) *The Ark on Ararat.* Thomas Nelson, Nashville.

Leakey, R.E. (1981) *The making of mankind.* Michael Joseph, London. 256 pp.

Lees, D.R. (1981) Industrial melanism: genetic adaptation of animals to air pollution. *In* Bishop, J.A. & Cook, L.M. (eds) *Genetic consequences of man made change.* Academic Press, London: 129–176.

Lewontin, R.C. (1974) *The genetic basis of evolutionary change.* Columbia University Press, New York. 340 pp.

Luria, S.E., Gould, S.J. & Singer, S. (1981) *A view of life.* The Benjamin/Cummings Publishing Co., Menlo Park.

Macdonald, G.A. & Abbott, A.J. (1970) *Volcanoes in the sea.* University of Hawaii, Honolulu.

Martin, L.D. (1984) Phyletic trends and evolutionary rates. *Special Publication Carnegie Museum of Natural History* No. 8: 526–538.

Mayr, E. (1982) *The growth of biological thought.* The Belknap Press of Harvard University Press, Cambridge, Mass. 974 pp.

McLoughlin, W.G. Jr (1955) *Billy Sunday was his real name.* Chicago University Press, Chicago. 324 pp.

McPhee, J. (1981) *Basin and Range.* Farrar, Straus & Giroux, New York.

Mellersh, H.E.L. (1968) Appreciation (of Charles Darwin). *In* Darwin, C. *The Voyage of the 'Beagle'.* Heron Books, London: 509–536.

Miller, H. (1857) *The Testimony of the Rocks.* Shepherd & Elliot, Edinburgh. 500 pp.

Miller, K. (1982) Answers to the standard Creationist arguments. *Creation/Evolution* 3: 1–13.

Miller, R. (1983) *Planet Earth — continents in collision.* Time-Life Books, Amsterdam. 176 pp.

Milne, D.H. & Schafersman, S.D. (1983) Dinosaur tracks, erosion marks and midnight chisel work (but no human footprints) in the Cretaceous limestone of the Paluxy River bed, Texas. *Journal of Geological Education* 31: 111–123.

Molnar, R. & Archer, M. (1984) Feeble and not so feeble flapping fliers: a consideration of early birds and bird-like reptiles. *In* Archer, M. & Clayton, G. (eds) *Vertebrate zoogeography and evolution in Australasia.* Hesperian Press, Perth: 407–419.

Montagu, A. (ed.) (1984) *Science and creationism.* Oxford University Press, New York. 415 pp.

Moore, R.A. (1983) The impossible voyage of Noah's Ark. *Creation/Evolution* 4: 1–43.

Morris, H.M. (1963) *The twilight of evolution.* Baker Book House, Grand Rapids, Mich. 103 pp.

Morris, H.M. (1967) *Evolution and the modern Christian.* Presbyterian & Reformed Publishing Co., Philadelphia.

Morris, H.M. (1972) *The remarkable birth of planet Earth.* Dimension Books, Minneapolis, Minn. 127 pp.

Morris, H.M. (ed.) (1974a) *Scientific creationism.* (General edition.) Creation-Life Publishers, San Diego. 277 pp.

Morris, H.M. (ed.) (1974b) *Scientific Creationism.* (Public school edition). Creation-Life Publishers, San Diego. 277 pp.

Morris, H.M. (1975) *The troubled waters of evolution.* Zonderman, San Diego. 75 pp.

Morris, H.M. (1977) *The scientific case for creation.* Creation-Life Publishers, San Diego.

Morris, J.D. (1980) *Tracking those incredible dinosaurs and the people who knew them.* Creation-Life Publishers, San Diego.

Morris, J.D. (1986) The Paluxy River mystery. *Impact* No. 151: i–iv.

Mortlock, R.P. (1982) Metabolic acquisitions through laboratory selection. *Annual Review of Microbiology* 36: 259–284.

Nei, M. & Koehn, R.K. (1983) *Evolution of genes and proteins.* Sinauer Associates Inc., Sunderland, Mass. 331 pp.

Neufeld, B. (1975) Dinosaur tracks and giant men. *Origins* 2(2): 64–76.

Ohno, S. (1984) Birth of a unique enzyme from an alternative reading frame of the preexisted, internally repetitious coding sequence. *Proceedings of the National Academy of Sciences, USA* 81: 2421–2425.

Orgel, L. (1982) Darwinism at the very beginning of life. *In* Cherfas, J. (ed.) *Darwin up to date.* IPC Magazines Ltd, London: 15–17.

Ostrom, J.H. (1979) Bird flight: how did it begin? *American Scientist* 67: 46–56.

Palmer, R.W. (1913) Note on the lower jaw and ear ossicles of a foetal *Perameles. Anatomischer Anzeiger Jena* 43: 510–515.

Patterson, C. (1978) *Evolution.* British Museum (Natural History), London & the University of Queensland, Brisbane. 197 pp.

Patterson, J.W. (1983) Thermodynamics and evolution. *In* Godfrey, L.R. (ed.) *Scientists confront creationism.* W.W. Norton, New York: 99–116.

Penny, D., Foulds, L.R. & Hendy, M.D. (1982) Testing the theory of evolution by comparing phylogenetic trees constructed from five different protein sequences. *Nature* (Lond.) 297: 197–200.

Popper, K.R. (1959) *The logic of scientific discovery,* Hutchinson, London.

Popper, K.R. (1978) Natural selection and the emergence of mind. *Dialectica* 32: 339–355.

Powell, J.R. (1978) The founder-flush speciation theory: an experimental approach. *Evolution* 32: 465–474.

Prakash, S. (1972) Origin of reproductive isolation in the absence of apparent genic differentiation in a geographic isolate of *Drosophila pseudoobscura. Genetics* 72: 143–155.

Rautian, A.S. (1978) A unique bird feather from Jurassic lake deposits in the Kara-tau Ridge Kasakh-SSR, USSR. *Paleontologicheskii Zhurnal* 4: 106–114.

Ride, W.D.L. (1985) Are Christians threatened by evolutionary biology? *St Mark's Review* No. 121: 10–22.

Rightmire, G.P. (1985) The tempo of change in the evolution of mid-Pleistocene *Homo. In* Delson, E. (ed.) *Ancestors: the hard evidence.* Alan R. Liss Inc., New York: 255–264.

Rogers, R. (1984) No faith in Creationism. *Life and Times,* June 13: 8.

Romer, A.S. & Parsons, J.S. (1977) *The vertebrate body.* 5th edition. W.B. Saunders, Philadelphia. 624 pp.

Ruse, M. (1982) *Darwinism defended.* Addison-Wesley, London. 356 pp.

Sagan, C. (1977) *The dragons of Eden.* Hodder & Stoughton, London. 263 pp.

Salvini-Plawen, L.V. & Mayr, E. (1977) On the evolution of photoreceptors and eyes. *In* Hecht, M.K., Steere, W.C. & Wallace, B. (eds) *Evolutionary biology,* Volume 10. Plenum Press, New York: 207–263.

Schadewald, R.J. (1982) Six 'Flood' arguments creationists can't answer. *Creation/Evolution* 3: 12–17.

Schadewald, R. (1985) The 1985 National Bible-Science Conference. *Creation/Evolution Newsletter* 5(4): 17–21.

Scott, I. (1980) History — influence and evolution, Part 3. *Ex Nihilo* 3(1): 21.

Setterfield, B. (1981) The velocity of light and the age of the universe. Part 1. *Ex Nihilo* 4 (1, 3, 4).

Siegler, H.R. (1978) A creationists' taxonomy. *Creation Research Society Quarterly* 15: 36–38.

Smith, D.G. (ed.) (1982a) *The Cambridge encyclopedia of Earth sciences.* Cambridge University Press, Cambridge. 496pp.

Smith, J.M. (ed.) (1982b) *Evolution now: a century after Darwin. Nature* & Macmillan, London. 239 pp.

Snelling, A., Mackay, J., Wieland, C. & Ham, K. (1983) The case against evolution: the case for creation. Casebook. *Ex Nihilo* 5(4): 13.

Soroka, L.G. & Nelson, C.L. (1983) Physical constraints on the Noachian deluge. *Journal of Geological Education* 31: 134–139.

Stanley, S.M. (1979) *Macroevolution, pattern and process*. W.H. Freeman, San Francisco. 332 pp.

Stansfield, W.D. (1977) *The science of evolution*. Macmillan Press Ltd, New York. 614 pp.

Stebbins, G.L. & Ayala, F.J. (1985) The evolution of Darwinism. *Scientific American* 253: 54–64.

Stevens, G.R. (1980) *New Zealand Adrift — The Theory of Continental Drift in a New Zealand Setting*. A.H. & A.W. Reed, Wellington. 442 pp.

Stringer, C.B. (1984) The definition of *Homo erectus* and the existence of the species in Africa and Europe. *In* Andrews, P. & Frazan, J.L. (eds) *The early evolution of man*. Courier Forschungsinstitut Senckenberg 69: 131–143.

Sullivan, W. (1974) *Continents in Motion: The New Earth Debate*. Macmillan, London. 399 pp.

Susman, R.L., Stern, J.T. Jr and Jungers, W.L. (1985) Locomotor adaptations in the Hadar hominids *in* Delson, E. (ed.) *Ancestors: The Hard Evidence*. Alan R. Liss., New York: 184–192.

Szalay, F.S. & Delson, E. (1979) *Evolutionary history of the primates*. Academic Press, New York. 580 pp.

Teilhard de Chardin, P. (1959) *The Phenomenon of Man*. Harper, New York. 318 pp.

Thulborn, R.A. (1985a) Birds as neotenous dinosaurs. *Records of the New Zealand Geological Society* 9: 90–92.

Thulborn, T. (1985b) Rot sets in in Queensland. *Nature* (Lond.) 315: 89.

Thulborn, T. (1985c) 'Gaps' in the fossil record. *In* Bridgestock, M. & Smith, K. (eds) '*Creationism': an Australian perspective*. Mark Plummer for the Australian Skeptics, Melbourne: 32–35.

Thulborn, T. (1985d) Pre-cambrian fossils. *In* Bridgestock, M. & Smith, K. (eds) '*Creationism': an Australian perspective*. Mark Plummer for the Australian Skeptics, Melbourne: 30–31.

Thwaites, W. & Awbrey, F. (1981) Biological evolution and the second law. *Creation/Evolution* 2: 5–7.

Tillich, P. (1955) *The new being*. Charles Scribner's Sons, New York. 179 pp.

Voorhees, R. (1985) Missouri round-table discussion on Biblical interpretation. *Creation/Evolution Newsletter* 5(3): 7–8.

Weaver, K.F. (1985) The search for our ancestors. *National Geographic* 168: 560–623.

Weber, C.G. (1981) Paluxy man — the creationist Piltdown. *Creation/Evolution* 2: 16–22.

Whetstone, K.N. (1983) Braincase of Mesozoic birds: I. New interpretation of the 'London' *Archaeopteryx*. *Journal of Vertebrate Paleontology* 2: 439–452.

Whitcomb, J.C. & Morris, H.M. (1961a) *The Genesis Flood*. The Presbyterian & Reformed Publishing Company, Nutley, New Jersey.

Whitcomb, J.C. & Morris, H.M. (1961b) *The Genesis Flood — The Biblical Record and its Scientific Implications*. Baker Book House, Grand Rapids, Mich. 518 pp.

White, M.J.D. (ed.) (1978) *Modes of Speciation*. W.H. Freeman, San Francisco.

Whitehead, A.N. (1978) *Process and reality*. Corrected edition, Griffin, D.R. & Sherburne, D.W. (eds). Free Press, New York. 413 pp.

Wielert, J.S. (1983) The creation–evolution debate as a model of issue polarization. *Journal of Geological Education* 31: 79–82.

Williams, M.B. (1982) The importance of prediction testing in evolutionary biology. *Erkenntnis* 17: 291–306.

Williamson, P.G. (1981a) Palaeontological documentation of speciation in Cenozoic molluscs from Turkana Basin. *Nature* (Lond.) 293: 437–443.

Williamson, P.G. (1981b) Morphological stasis and developmental constraint: real problems for neo-Darwinism. *Nature* (Lond.) 294: 214–215.

Wilson, A.C. (1985) The molecular basis of evolution. *Scientific American* 253(4): 148–157.

Wolpoff, M.H. (1984) Evolution of *Homo erectus*: the question of stasis. *Paleobiology* 10: 389–406.

Wood, R.J. & Bishop, J.A. (1981) Insecticide resistance: populations and evolution. *In* Bishop, J.A. & Cook, L.M. *Genetic consequences of man made change*. Academic Press, London: 97–127.

Yockey, W.P. (1977) A calculation of the probability of spontaneous biogenesis by information theory. *Journal of Theoretical Biology* 67: 377–398.